Butterfly Conservation in North America

AF514202

Jaret C. Daniels
Editor

Butterfly Conservation in North America

Efforts to help save our charismatic microfauna

Springer

Editor
Jaret C. Daniels
McGuire Center for Lepidoptera and Biodiversity
Florida Museum of Natural History
and Department of Entomology
and Nematology
University of Florida
Gainesville, FL, USA

ISBN 978-94-017-7935-7 ISBN 978-94-017-9852-5 (eBook)
DOI 10.1007/978-94-017-9852-5

Springer Dordrecht Heidelberg New York London
© Springer Science+Business Media B.V. 2015
Softcover reprint of the hardcover 1st edition 2015
This work is subject to copyright. All rights are reserved by the Publisher, whether the whole or part of the material is concerned, specifically the rights of translation, reprinting, reuse of illustrations, recitation, broadcasting, reproduction on microfilms or in any other physical way, and transmission or information storage and retrieval, electronic adaptation, computer software, or by similar or dissimilar methodology now known or hereafter developed.
The use of general descriptive names, registered names, trademarks, service marks, etc. in this publication does not imply, even in the absence of a specific statement, that such names are exempt from the relevant protective laws and regulations and therefore free for general use.
The publisher, the authors and the editors are safe to assume that the advice and information in this book are believed to be true and accurate at the date of publication. Neither the publisher nor the authors or the editors give a warranty, express or implied, with respect to the material contained herein or for any errors or omissions that may have been made.

Printed on acid-free paper

Springer Science+Business Media B.V. Dordrecht is part of Springer Science+Business Media (www.springer.com)

Acknowledgments

The initial concept for this book resulted from a series of national workshops focused on imperiled butterfly conservation and management that took place over a 3-year period. Funded by the Institute of Museum and Library Services (MP-00-08-0022-08), the intensive cross-training program targeted conservation practitioners from zoos, botanical gardens, natural history museums, nature centers, state and federal wildlife agencies, and academic institutions. Over the several years it took to prepare the resulting manuscript, a great many individuals have helped shape the final product. I am grateful for the tolerance of my colleagues and chapter authors, to Springer Science + Business Media B.V. for their patience and assistance, and to the countless conservation practitioners who have made a significant difference for the recovery of the butterfly taxa highlighted. I additionally want to thank Geena Hill and Stephanie Sanchez for their assistance in reviewing chapter versions and providing key editorial assistance.

The following authors and organizations are thanked for the permission to use images and/or reproduce or modify diagrams: Amber Hasselbring (Nature in the City), Danielle LeFer (Palos Verdes Land Conservancy), Melissa McGaw, and Chris van Swaay.

Contents

Butterfly Recovery Planning: Determining How to Contribute

Jaret C. Daniels, Mitch Magdich, and Peter Tolson

Abstract Interest in at-risk butterfly conservation has grown tremendously in recent years, as has the number of dedicated recovery initiatives. Zoos, natural history museums, botanical gardens, and state and federal wildlife agencies are progressively focusing on insects, particularly charismatic groups such as butterflies and pollinators, to help advance local conservation efforts and foster increased public interest and community engagement. However, insufficient experience and familiarity with butterflies can often hinder conservation practitioners from adequately planning, implementing and evaluating essential program components. Determining the best ways to make meaningful contributions to new or ongoing at-risk butterfly recovery initiatives is critical and typically driven by available expertise and resources. This book is intended to serve as a basic primer for practitioners interested in working with butterflies. The various chapters provide a combination of specific case studies and broader overviews of key themes inherent to most initiatives. A detailed decision tree is discussed to review available options.

Introduction

Butterflies are experiencing significant declines worldwide. Habitat loss, habitat degradation, and climate change are arguably the primary global drivers responsible for the erosion and extinction of populations at various scales (van Swaay et al. 2013; Fox et al. 2006; Thomas et al. 2004; Thomas and Clarke 2004). In the United States, there are currently 29 butterflies federally listed as threatened or endangered under the Endangered Species Act (ESA). There are three that are federally-designated as threatened because of similarity of appearance to endangered taxa (US Fish and Wildlife Service 2014). Many more receive protection at the

J.C. Daniels (✉)
McGuire Center for Lepidoptera and Biodiversity, Florida Museum of Natural History and Department of Entomology and Nematology, University of Florida, Gainesville, FL, USA
e-mail: jdaniels@flmnh.ufl.edu

M. Magdich • P. Tolson
The Toledo Zoo, Toledo, OH, USA
e-mail: mitchell.magdich@Toledozoo.org; peter.tolson@toledozoo.org

© Springer Science+Business Media B.V. 2015
J.C. Daniels (ed.), *Butterfly Conservation in North America*,
DOI 10.1007/978-94-017-9852-5_1

individual state level or which are regarded as species of special concern or greatest conservation need. Canada lists twelve butterflies as threatened or endangered and three special concern species under the Species at Risk Act (SARA) (Species at Risk Public Registry 2014), with an additional three species officially designated as extirpated. Even iconic, once abundant butterflies like the monarch have shown dramatic losses in recent years prompting significant concern by conservationists (Flockhart et al. 2014).

While official listing confers many benefits on the affected species, listing status is a poor proxy for assessing overall degree of imperilment, especially for underrepresented groups such as insects. For example, despite a rich subtropical fauna of more than 200 recorded species, including many endemic or habitat-specialist taxa, only seven Florida butterflies receive legal protection at the state or federal level. Of these, two represent recent 2014 additions to the ESA (US Fish and Wildlife Service 2014) and another three are federally-designated as threatened because of geographic overlap and similarity of appearance to an endangered taxon. By contrast, the Florida Natural Areas Inventory (FNAI) currently tracks 69 at-risk butterfly taxa. As one of 82 natural heritage programs operating in North and Latin America under the NatureServe umbrella (NatureServe 2014), the FNAI maintains a statewide database on the locations and condition of federally- or state-listed plants, animals, and ecological communities as well as those that are deemed to be at risk of becoming rare, threatened, or endangered within Florida in the near future.

As the rate of species imperilment continues to accelerate, increased conservation actions, including more proactive, innovative and collaborative approaches to promote organism recovery, are necessary. Many conservation organizations and informal science education institutions such as zoos, natural history museums and botanical gardens are progressively focusing on insects, particularly charismatic groups such as butterflies and native pollinators, to help advance local conservation efforts and foster increased community interest and engagement. However, insufficient experience and familiarity with insects is often a critical constraint preventing staff and institutions from adequately planning, implementing and evaluating specific conservation or recovery actions. Similarly, few biologists working for wildlife agencies at the federal, territorial or state level have a strong entomological background. In many cases, the development of targeted recovery or action plans designed to address the specific projects or activities required to meet the goals and objectives outlined in an overall recovery strategy are overseen by agency staffs that traditionally have much more extensive vertebrate or botanical expertise. The resulting knowledge gaps can be considerable, specifically as they relate to issues of land management, habitat restoration, ecology and population dynamics, population monitoring, and captive breeding. Consequently, practitioner training and institutional assistance are needed to facilitate more comprehensive and integrated approaches to butterfly conservation where ex situ efforts are part of the broader toolbox used to support in situ conservation of wild populations and their habitats. This is particularly true with the recent proliferation of new at-risk butterfly conservation programs that often involve a growing list of nontraditional partners. For many parties, just knowing where to begin or how to effectively contribute is often half the battle.

This book is intended to serve as a basic primer for conservation practitioners interested in helping to plan and implement various program components as they relate to at-risk butterfly recovery. The various chapters provide a combination of specific case studies and broader initial overviews of key themes inherent to most initiatives. Comprehensive detail of these and other related topics are readily available in numerous other standard texts and would serve as valuable additional reading.

Learning from the Network: Promoting Enhanced Cross-Program Communication and Standardization

While the field of insect conservation continues to develop rapidly, it remains an inexact science. Incomplete data, particularly related to species-specific biology and sensitivities to particular threats or habitat changes, retard informed decision-making and regularly hinder the implementation of practical conservation and management. As a result, most at-risk butterfly conservation initiatives are inherently pioneering. Nonetheless, overall interest in butterfly conservation has grown tremendously in recent years, as has the number of dedicated programs. The end results of such efforts have often been mixed. Examples exist of failed programs as well as those that have slowed the decline of some species and even led to the recovery or resurgence of others (Thomas et al. 2009). Limitations of their conservation effectiveness may arguably be tied as much to a lack of integration and standardization across programs as they are to the biological complexities of the target organism or system. Sutherland et al. (2010) suggest that "standardizing methods and making the information freely available will greatly facilitate the improvement of conservation practice through enabling others in the conservation community to learn from the experience of each contributing individual and organization." They go on to recommend a minimum set of standards for documenting and monitoring program methods and outcomes. Although developed specifically for avian reintroduction projects, the basic standards outlined are an excellent addition to the conservation literature. They could readily be adapted for use with a variety of taxa including butterflies or serve as an initial template to help catalyze the development of similar standards appropriate to other conservation and recovery program components, especially where there is variation in techniques.

In a review of British and American at-risk butterfly conservation efforts, Schultz et al. (2008) echo the need for more systematic communication and record keeping across programs. Improved data standardization and sharing of lessons learned and best practices would facilitate effective programmatic evaluation, cross program reviews and the advancement of new or existing programs. The authors additionally stress the need for "improved experimental design and enhanced monitoring to improve our understanding of the mechanisms by which habitat management influences the general ecology and population trends of at-risk species." The ability to more effectively connect such on-the-ground efforts to broader ecological responses of species would enable more strategic allocation of limited resources and significantly improve the implementation and impact of conservation actions.

Basic Components of At-Risk Butterfly Conservation Strategies

The International Union for Conservation of Nature (IUCN) asserts that "a goal of conservation is the maintenance of existing genetic diversity and viable populations of all taxa in the wild in order to maintain biological interactions, ecological processes and function" (IUCN 2002). IUCN further acknowledges that "it will not be possible to ensure the survival of an increasing number of threatened taxa without effectively using a diverse range of complementary approaches and techniques." As with many other taxa, the conservation strategies to help recover at-risk butterflies typically involve a range of in situ and ex situ actions including threat mitigation, habitat restoration, enhancement and management, captive propagation, organism reintroduction or translocation, monitoring, and stakeholder education and training (New 1991; Schultz et al. 2008).

Laboratory and Field Research

An appropriate research-based approach to organism conservation is critical to help advance organism recovery. Combined with modern planning and evaluations tools such as the Open Standards for the Practice of Conservation (Salafsky et al. 2002; Conservation Measures Partnership 2004), conservation practitioners can maximize the effectiveness and efficiency of their efforts through the implementation of an adaptive management cycle. At the programmatic level however, recovery planning for at-risk species is often hindered by a fundamental lack of information on many fronts. Research that specifically targets key data gaps should be prioritized. Topics to be emphasized include but are not limited to:

- Reasons for decline
- Current threats and threat mitigation strategies
- Basic biology, ecology and behavior
- Organism husbandry
- Population genetics, including issues of taxonomic resolution and phylogenetic relationships, population genetic structure, genetic health, minimum viable population size and frequency of gene flow/connectivity of populations (see chapter "Butterfly Conservation Genetics" for more comprehensive overview)
- Suitable habitat requirements
- Habitat assessment and mapping
- Construction of population viability assessments/analyses and other population modeling

- Organism responses to habitat enhancements, restoration and/or management actions
- Best management practices (see chapter "Managing Land for Butterflies" for a more comprehensive overview)
- Reintroduction or augmentation methods and efforts
- Population assessment and monitoring (see chapter "Butterfly Monitoring for Conservation" for a more comprehensive overview)

For all targeted research projects, extensive opportunities exist for collaboration with wildlife agencies, conservation organizations, land managers, colleges and universities and even citizen scientists.

Habitat Restoration and Enhancement

For a great many at-risk species, the lack of suitable habitat is the major barrier to recovery. Efforts to help restore, enhance or even establish habitat upon which at-risk species populations depend and the development of best management practices to help ensure the long-term maintenance of those populations and critical resources are essential components of virtually all conservation or recovery plans.

The terminology used here to define these specific actions generally follows the National Research Council (1992):

Habitat Restoration is the "manipulation of the physical, chemical, or biological characteristics of a site with the goal of returning natural/historic functions to former or degraded" habitats.

Habitat Enhancement is the "manipulation of the physical, chemical, or biological characteristics of a site to heighten, intensify, or improve specific function(s)."

Habitat Establishment is the "manipulation of the physical, chemical, or biological characteristics present to support and maintain habitat that did not previously exist" on the site.

Such landscape-level efforts are inherently labor intensive and often involve extended time scales to yield desired results. They can involve a diverse array of actions including invasive species control, planting native grasses and forbs (especially larval host plants and adult nectar sources), enhancing structural diversity, and restoring or more appropriately implementing processes like fire to benefit the target species. As a result, they generally offer many entry points for contribution and collaboration. Such programmatic components can be ideal for conservation organizations such as zoos, botanical gardens, natural history museums and nature centers that can offer specific expertise, have access to volunteer or member pools to help provide much needed on-the-ground assistance, and have broad environmental education and conservation missions. Outcomes often extend beyond improving ecosystem function and habitat quality or availability for a broad array of organisms. Initiatives of this nature can also promote environmental literacy, engender public conservation support and empower local communities.

Captive Propagation

The use of *ex-situ* conservation efforts defined by the Convention on Biological Diversity (see Glowka et al. 1994) as "the conservation of components of biological diversity outside their natural habitats" has grown tremendously in recent decades. The bulk of this increase has been in zoos, aquariums, natural history museums, botanic gardens and related research facilities which provide combined facilities, intellectual investment to conservation, and subject expertise (often in areas of research, husbandry or restoration). In their review of British and American butterfly conservation efforts, Schultz et al. (2008) highlight that captive propagation is recommended in the conservation strategies for nearly half of U.S. species listed as Endangered, Threatened, or Candidate under the US Endangered Species Act. That list has grown in recent years with additional listings. With the proliferation of new programs has also been a basic shift in the traditional 'millennium ark' paradigm suggested by Soulé et al. (1986). The long-term maintenance of self-sustaining captive populations that can withstand demographic and environmental stochasticity while maintaining evolutionary potential has moved to a more comprehensive and integrated approach where ex situ efforts are part of the broader practitioner toolbox used to support in situ conservation of wild populations and their habitats. The rapidly advancing array of sophisticated scientific tools and protocols available to programs has additionally made distinctions between such species conservation practices less pronounced.

Nonetheless, the use of ex situ conservation practices remains controversial and has numerous practical and fundamental limitations (Snyder et al. 1996). Compared to larger vertebrates or many aquatic organisms, efforts targeting butterflies or other insects can often present reduced challenges particularly in terms of space, costs, sophistication of facilities, and complexity of husbandry techniques. In all cases however, the end goal is the same sustainability (albeit often very time-limited) of healthy, genetically diverse, and demographically varied populations that can (in some manner) enhance the conservation of wild populations (AZA 2012). Inherent in this process should be the development of appropriate protocols and measures to minimize adaptation to captive conditions, and thus potentially hinder reintroduction success.

Within the context of this publication, captive propagation serves as a broader umbrella that encompasses a variety of similar captive measures including head-starting, captive rearing and captive breeding. For purposes of clarity, the following definitions are provided:

Head-Starting: raising immature (early) stage organisms that were collected in the wild to a later life stage in captivity before releasing them back into the wild in a suitable habitat (adapted from Smith and Sutherland 2014). As the term implies, this technique is mean to provide enhanced organism survival or "a helping hand" under very short-term captive conditions before eventual release. It has been a successful *ex situ* component that is currently most associated with amphibian conservation and recovery efforts. In relation to butterflies, it could involve the collection of eggs or young larvae from the wild and the release of more mature larvae, pupae or even adults.

Captive Rearing: similar to the above but typically involves the completion of a full generation in captivity. No breeding of individuals is undertaken and the captive population is not self-sustaining. In relation to butterflies, this could involve the collection of wild adults and the rearing to maturity of their offspring.

Captive Breeding (or Captive Propagation): involves the establishment of a self-sustaining population in captivity for a defined period. Multiple generations are typically produced.

The use of such ex situ conservation practices should only be implemented as part of a broader approved recovery plan or strategy and in association with other integrated in situ conservation measures. It should be noted that while captive breeding or rearing efforts are often conducted for the purpose of providing organisms for reintroduction, augmentation or introduction efforts, this is but one possible goal of such programs. Other supported outcomes include maintaining refugia populations, supporting recovery-related research, and conserving species or populations that are in imminent risk of extinction or extirpation (US Fish and Wildlife Service 2000). Therefore, clearly defined and articulated programmatic goals should be determined with the appropriate stakeholder community prior to initiation of such intensive recovery efforts.

Education

Education is often used as a broad, catch-all term. For purposes relating to butterfly recovery, it is meant to encompass public conservation education and stakeholder training. While often not catapulting to the top of the list as a key priority, education and training are almost always elements included in comprehensive recovery plans or strategies. Such components are essential to help promote increased awareness and knowledge of at-risk species, their habitats and the various threats affecting them. They can also play an important role in enhancing environmental literacy, stimulating increased connection to nature, fostering positive attitudes toward conservation, and promoting careers in conservation-related disciplines. Informal science education organizations such as zoos, aquariums, botanical gardens, natural history museums, and nature centers reach hundreds of millions of people each year. Such institutions are particularly adept at communicating often complex conservation messages to diverse audiences in an assortment of ways to help maximize learning outcomes. These may include traditional print and electronic resources, outreach, camps, programming, exhibits, festivals, citizen science initiatives, curricula, hand-held technology, gaming, and even theatre.

Due to their broad public appeal and utility as conservation flagships, butterflies can be relatively simple but strategic additions to new or existing educational programming. In addition, they offer numerous opportunities to effectively integrate and interpret both basic and complex environmental concepts ranging from ecosystem services and evolution to climate change and diversity.

Stakeholder training is an equally important component that can significantly benefit planning, implementing and evaluating recovery-related activities. Expanded training opportunities may be particularly valuable to wildlife agency biologists and land managers to learn new techniques and best practices, forge additional ties to academic and professional communities, and receive guidance on emerging issues (Crouse et al. 2002). Even simple tasks such as informing land managers about the at-risk species on their property and where critical habitat occurs can be highly impactful. In the process, such interactions additionally foster increased stakeholder communication, collaboration, and capacity. The resulting growth in diversity of contributing stakeholders and scientific participation benefits the overall understanding, implementation and acceptance of recovery plans (Tolson and Magdich 2001; Crouse et al. 2002; Daniels 2009).

Getting Involved with Butterfly Conservation: Considering Your Options

The complexity of program component options and array of corresponding scientific tools and protocols available can be burdening to conservation organizations and practitioners. Determining the best ways to make meaningful contributions to new or ongoing at-risk butterfly recovery initiatives is critical and typically driven by available expertise and resources. To help facilitate this process, a decision tree is often a useful decision support tool that can be employed to review available options. Figure 1 and the following outline provide a broad perspective that can be used to help select potential entry points. It is essential to be realistic during this initial phase. Conduct a detail assessment of your organization's available resources, staff expertise and overall level of administrative commitment before considering any potential option.

(A) There are generally five limiting factors that come into play when considering your organization's level of involvement in at-risk butterfly recovery.

1. Available funds
2. Available time
3. Available space
4. Available personnel
5. Staff expertise

(B) All of these factors, in combination, will help you determine the extent of your involvement.

1. Low resources

    (a) Very limited access to funds, time, space, personnel or staff expertise

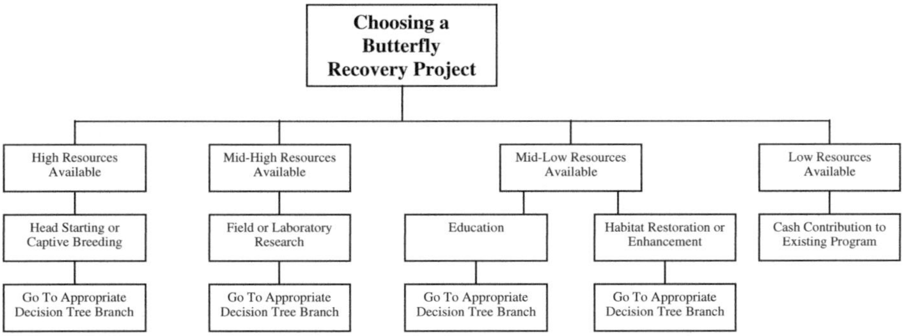

Fig. 1 Decision tree for choosing a butterfly recovery project based on available resources

2. Low-mid resources
   (b) Limited access to funds, time, space, personnel or staff expertise
3. Mid-high resources
   (c) Adequate access to funds, time, space, personnel or staff expertise
4. High resources
   (d) Ample access to funds, time, space, personnel or staff expertise

(C) After determining where your institution is on the continuum, you can make a more informed decision regarding your level of involvement.

1. Low resources
   (a) Financial contribution to existing recovery program
2. Low-mid resources
   (a) Education component
   (b) Habitat enhancement or restoration component
3. Mid-high resources
   (a) Field or laboratory research component (includes population surveys or monitoring efforts)
4. High resources
   (a) Head starting or captive breeding component

Once your initial review is complete and potential options have been tentatively identified, a more exhaustive analysis is necessary to determine the best course of action. The following detailed outline and corresponding tailored decision tree branches can be used to further explore various strategy pathways, identify key questions and requirements, and help launch initial discussions with potential partners or stakeholders.

(A) **Financial contribution** to existing program

1. Quickest and easiest way to help support at-risk butterfly conservation
2. Often ideal for smaller organizations/institutions that have limited resources, especially in areas of time, personnel and expertise
3. Even a limited financial contribution can have a significant impact

(B) **Education component** (Fig. 2)

1. Depending on the scope, an educational component can be initiated without extensive funding, time, expertise or personnel requirements
   (a) The resource requirements will increase depending on the scope, time-frame and complexity of the project
   (b) Can provide needed visibility to help enhance public awareness and understanding of at-risk butterflies and the habitats that sustain them
2. Involvement could include:
   (a) Educational programming
   (b) Curriculum development
   (c) Teacher training
   (d) Exhibit development
   (e) Special events
   (f) Development of print or electronic support materials (brochures, lesson plans, fact sheets, discovery kits, etc.)
3. Many new educational components can be effectively integrated into preexisting programs such as camps, events, classes, and outreach.
4. Criteria
   (a) Secure buy in from your educational staff or volunteer staff
   (b) Identify a target project and its overall scope
   (c) Identify the target audience(s)
   (d) Determine the implementation date and duration of the project
      1. Are additional financial resources required?
   (e) Determine if the appropriate expertise is available to implement the project
      1. If additional expertise is required, what opportunities for collaboration exist?
   (f) Identify and acquire the needed materials to implement and sustain the program
      1. Is specialized equipment required?
   (g) Determine all project deliverables

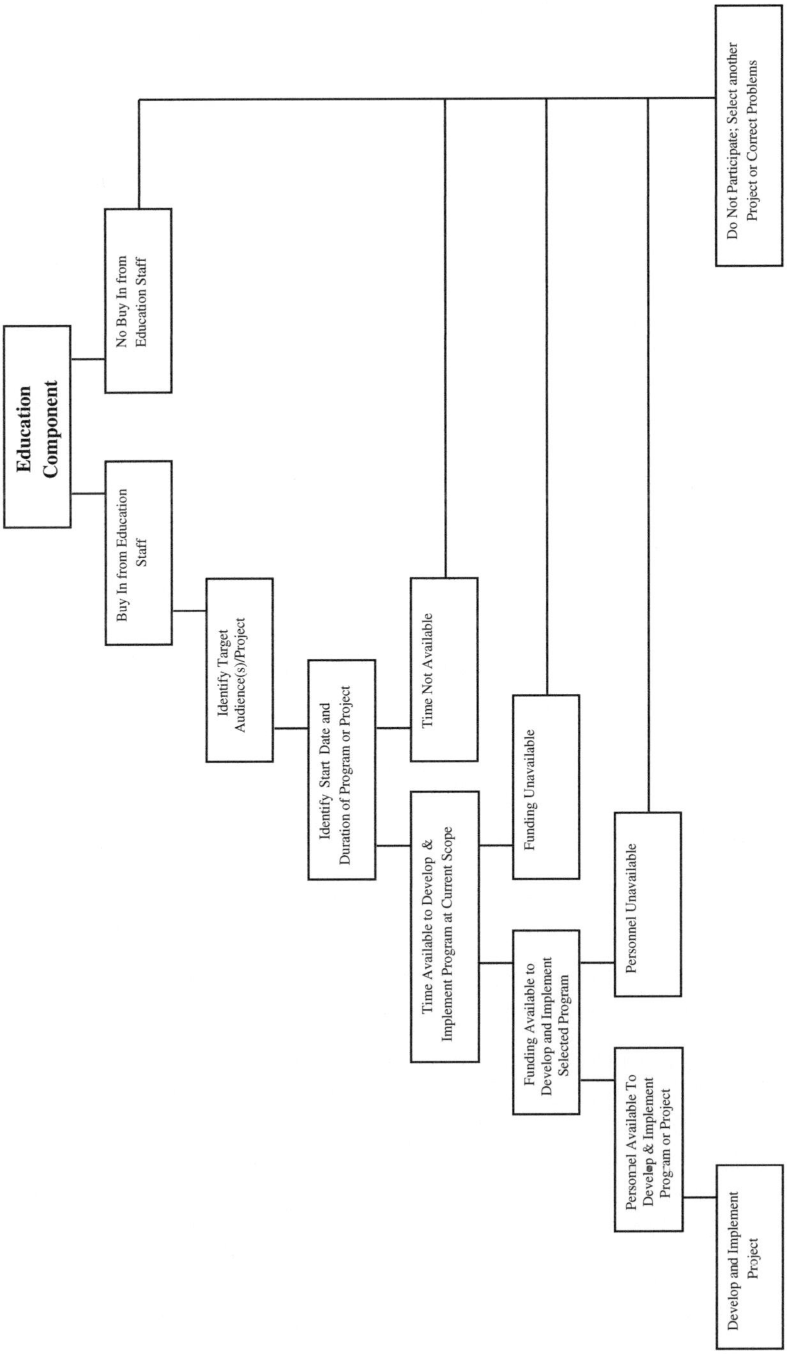

Fig. 2 Educational component decision tree branch

(h) Develop a dissemination plan for to ensure deliverables reach identified target audience(s) and other key stakeholders if applicable
(i) Develop and implement an evaluation plan to determine the effectiveness of the program component

(C) **Habitat enhancement or restoration component** (Fig. 3)

1. Participation would include collaboration with other conservation organizations or agencies by supplying paid staff, volunteers or materials to assist in habitat enhancement or restoration projects.
2. Institutional commitment would vary depending on the size, scope and duration of the project.

   (a) Potential levels of involvement

      1. Providing labor to assist with habitat restoration or management (mowing, prescribed fire, invasive species removal, etc.)
      2. Providing abor to assist with habitat enhancement (re-vegetation, planning target larval host plants and adult nectar sources, etc.)
      3. Providing labor, space and expertise to propagate plant material for re-vegetation or augmentation efforts
      4. Providing labor and expertise to conduct or assist with habitat assessments

3. Criteria

   (a) Identify target project (habitat, location, etc.)
   (b) Identify lead agencies, landowners and potential partners
   (c) Determine level and duration of involvement
   (d) Coordinate with lead agencies, landowners and partners regarding planning, logistics and timing of all activities (includes acquiring all necessary permits, securing access to site and any necessary supplies or equipment, determining safety and liability issues, etc.)
   (e) Determine if there are sufficient resources available to ensure consistent participation

(D) **Field or laboratory research component** (Fig. 4)

1. The field or laboratory research component will generally be more resource intensive than the previous program components, depending on the scope, scale and duration of the research.
2. Projects could involve any aspect that helps advance knowledge related to the management and recovery of the target organism or habitat (organism biology, ecology and behavior, population monitoring, habitat requirements, threat mitigation, habitat management or restoration, etc.)
3. Projects could be as simple as a presence/absence survey or as complex as a long-term ecological study.
4. Criteria

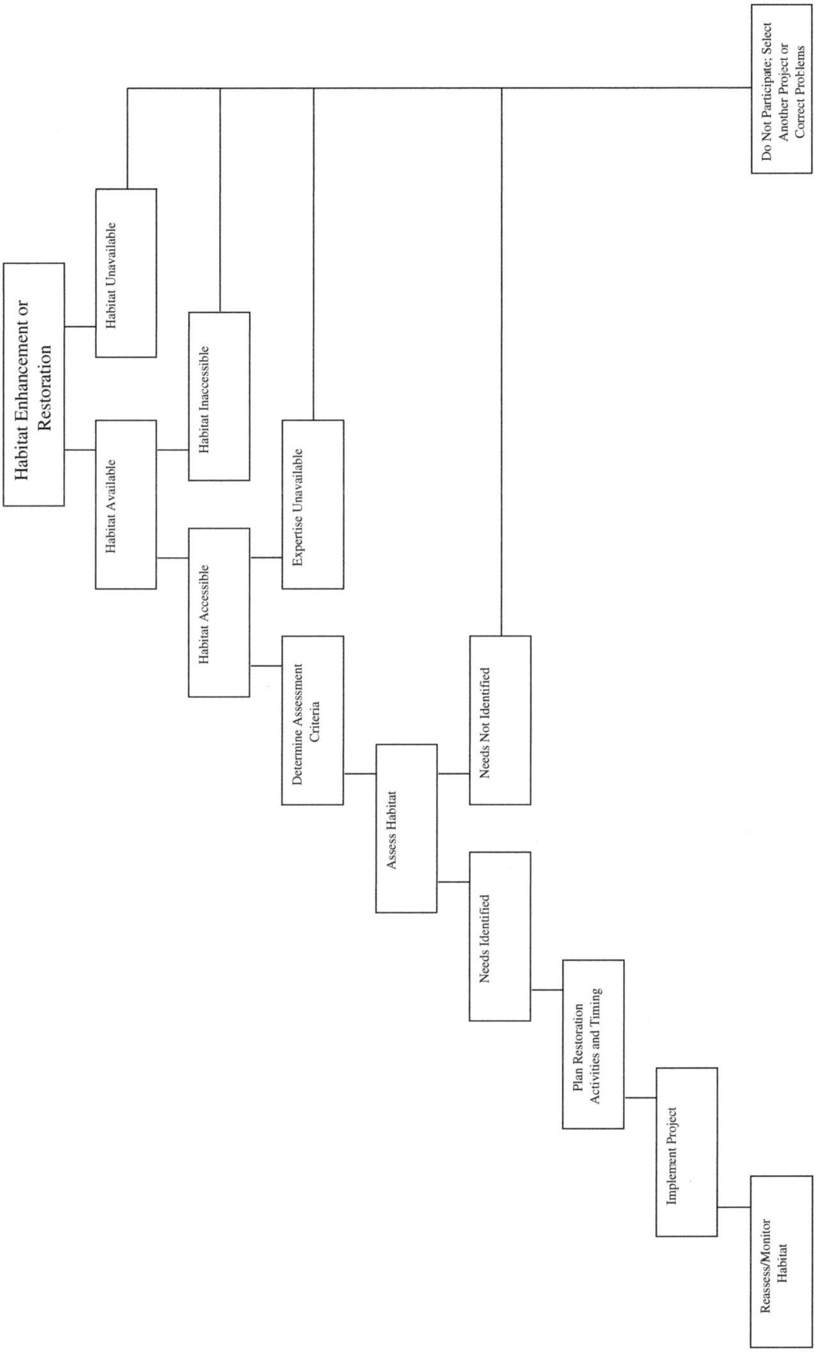

Fig. 3 Habitat restoration or enhancement recovery component decision tree branch

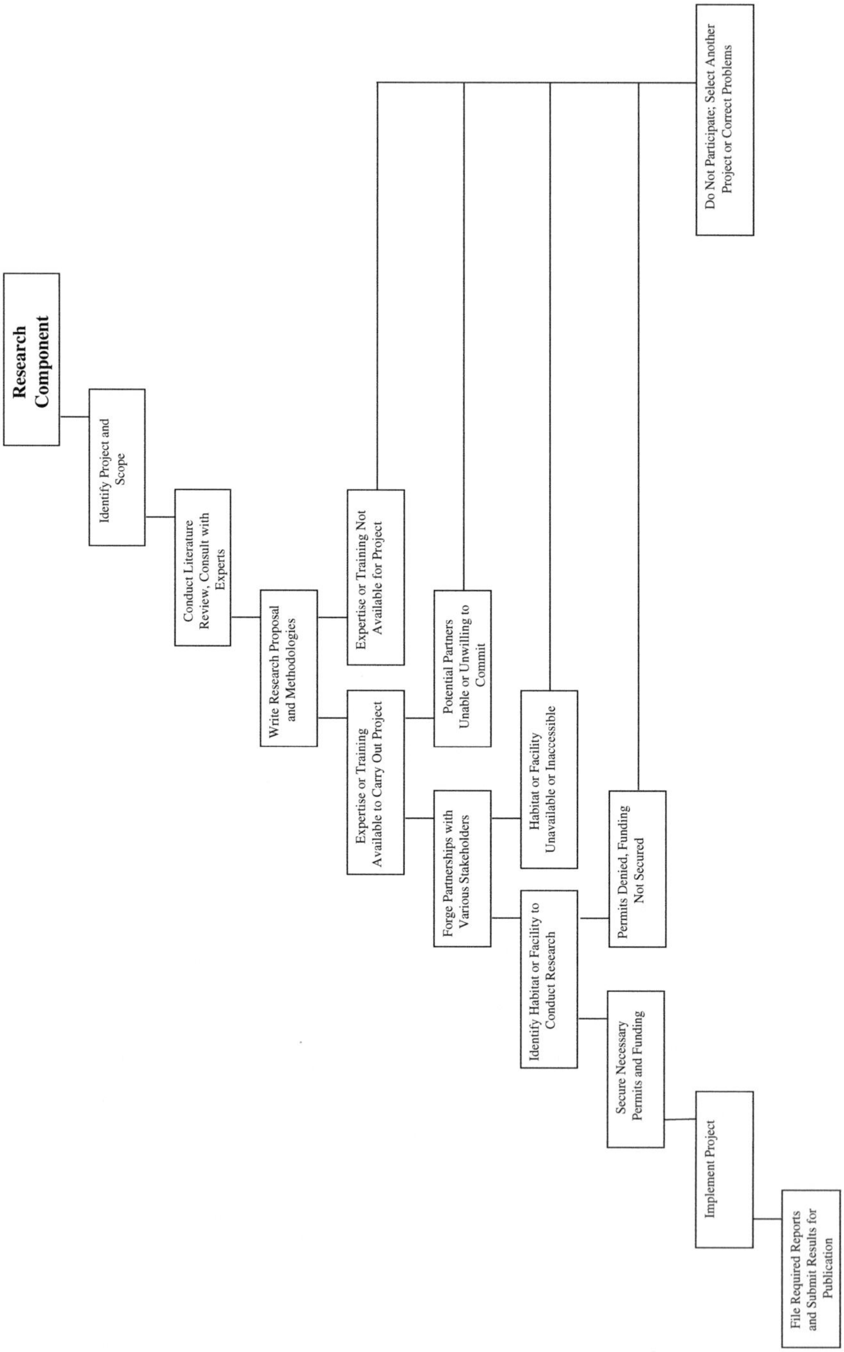

Fig. 4 Field or laboratory research recovery component decision tree branch

(a) Identify target organism
(b) Assess key research needs by reviewing available conservation, recovery or management plans and consulting with lead agencies or experts
(c) Identify the project and the scope based on your institution's available resources
(d) Determine if additional resources are required
(e) Conduct a thorough literature review and consult with lead agencies or experts to help determine an appropriate research plan and experimental design
(f) Develop a detailed research proposal
(g) Forge partnerships with key stakeholders such as landowners, NGO's, governmental agencies and universities
(h) Secure all necessary permits or permissions needed to conduct the research
(i) Implement the project, collect and analyze the data, and file the necessary reports
(j) Submit findings for publication

(E) **Captive breeding component** (Fig. 5a, b)

1. This is probably the most resource intensive component since it may combine elements of other previous components (e.g., research, habitat enhancement and restoration)
2. There are numerous criteria that need to be addressed:

   (a) Identify the target organism and determine the scope of the project

      1. What is the conservation status of the target organism?
      2. What permits are required and are they obtainable?
      3. What are the established goals of the captive program (research, population augmentation, assurance population, etc.)?
      4. Conduct a literature review and consult with lead agencies or experts

   (b) Review the biology of the target organism

      1. Is the biology of the butterfly well understood or will there be a need for additional research?
      2. Does the butterfly have complex life history requirements that would complicate captive breeding?
      3. What are the larval host plants?
      4. Can the larval host plants be readily attained, propagated and maintained?
      5. What are the nutritional requirements of the adult?
      6. How many generations per year does the butterfly have?
      7. In what life stage does the butterfly diapause?
      8. Will adults mate in captivity?

   (c) Forge partnerships with key stakeholders in the project

a

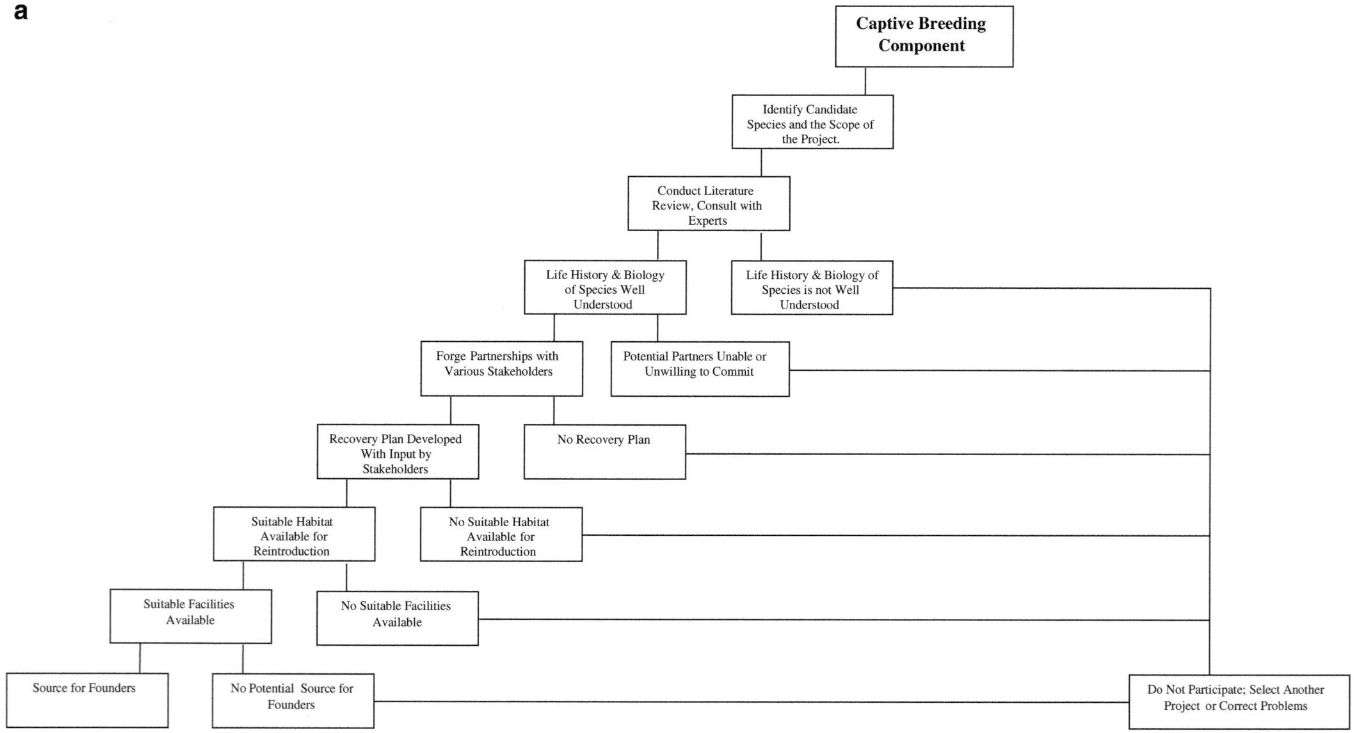

Fig. 5 (**a**) Captive breeding recovery component decision tree branch. (**b**) Captive breeding recovery component decision tree branch continued

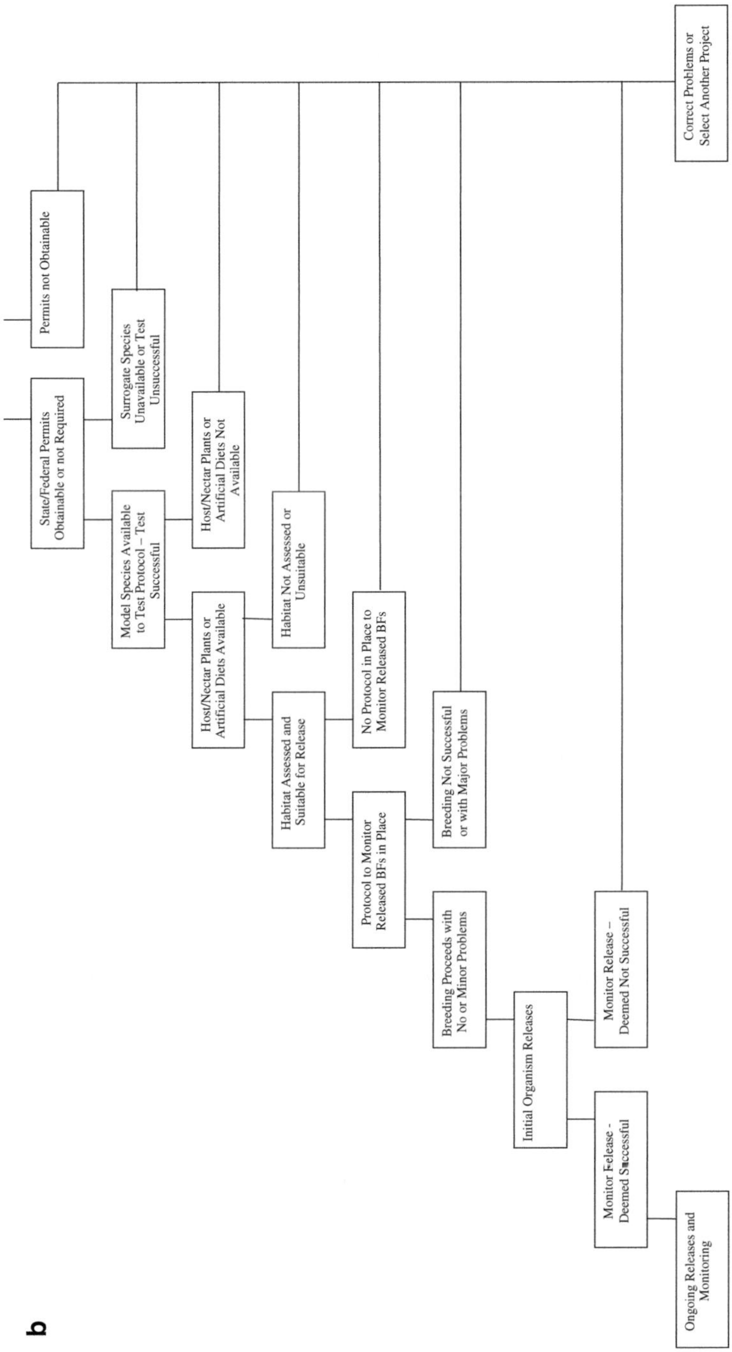

Fig. 5 (continued)

(d) Join an existing recovery team or working group (or collaborate with stakeholders to initiate one)

1. What is the status of the overall recovery effort?
2. Does a captive breeding plan exist?
3. If so, what is the status (incomplete, draft, final)
4. Is there an existing captive breeding program?
5. What established protocols exist?
6. Can protocols be adapted or modified?

(e) Identify a suitable facility to conduct the captive breeding

1. What life stage(s) will be maintained in captivity?
2. What are the environmental requirements (light, temperature, space, etc.) of the targeted life stage(s)?
3. What are the space requirements of the program?
4. What support components are required to effectively maintain the program (e.g. greenhouse, plant nursery)
5. Is the facility secure and equipped with appropriate containment to keep organisms from escaping?

(f) Determine a realistic annual operating budget

1. Identify any required new construction or existing facility renovation
2. Identify required support personnel and associate costs for daily maintenance of livestock, plants, facilities and all other associated program support efforts.
3. Identify required support materials (cages, plants, nursery supplies, rearing containers, etc.) and all related acquisition costs

(g) Address the nutritional requirements of the larvae and adults.

1. Will larval host plants and adult nectar plants need to be propagated?
2. If so, what are the necessary requirements for propagation and maintenance?
3. Are particular plant parts or conditions required for oviposition or larval feeding (e.g. flowers, fruit, new growth, age or height class)
4. Can an artificial diet be used for either or both the larvae and adults?

(h) Identify a source of founders

1. Are founders readily available?
2. Are founders genetically compatible with the current population if the project is an augmentation?
3. What is the status of the founder population?

(i) Develop a data collection and management plan

1. Comprehensive data collection and management is critical to help evaluate the program success

2. Regularly evaluate all efforts
3. Make necessary changes and continue

(j) Develop a detailed proposal for the targeted effort in coordination with key agencies and recovery team members

1. If additional funding is required, identify potential funding sources and submit proposal

(k) Apply for necessary permits
(l) Identify a surrogate species to test the proposed protocol

1. Is there a secure and closely related species that can be used?
2. Is the surrogate readily available?
3. Are the life history requirements similar (host use, number of generations, behavior)?
4. Are permits required to capture founders?

(m) Work with recovery team to identify suitable habitat for potential release sites

1. Is the habitat sufficient to support a release?
2. With vegetation surveys be required to ascertain the suitability of the habitat?
3. Will any restoration be required?
4. Are resources available to restore the habitat if necessary?

(n) After habitat restoration is complete, who will conduct the assessment to determine the readiness of the site for organism reintroductions?
(o) Work with the recovery team to determine the release requirements

1. What life stage(s) will be released?
2. How many organisms will be released and when?
3. When will releases be conducted?
4. Are multiple releases going to be conducted?

(p) Develop a protocol for monitoring the released organisms
(q) Proceed with capture and initial breeding trials using the surrogate species
(r) Proceed with capture and breeding of target at-risk species (once the protocol has been evaluated and surrogate breeding has been considered successful)
(s) Proceed with initial organism release(s)
(t) Monitor the release(s)
(u) Continue with breeding, organism releases and monitoring in successive seasons as determined by the recovery team
(v) Submit all required reports
(w) Goal of the project has been met

Summary and Key Points

- Butterflies are experiencing significant declines worldwide. As the rate of species imperilment continues to accelerate, increased proactive, innovative and collaborative conservation actions are necessary.
- Many conservation practitioners are progressively focusing on insects, particularly charismatic groups such as butterflies and native pollinators, to help advance local conservation efforts and foster increased community interest and engagement. However, insufficient experience and familiarity with insects is a critical constraint preventing staff and institutions from adequately planning, implementing and evaluating recovery actions.
- Conservation strategies to help recover at-risk butterflies typically involve a range of in situ and ex situ actions including threat mitigation, habitat restoration, enhancement and management, captive propagation, organism reintroduction or translocation, monitoring, and stakeholder education and training.
- Determining the best ways to make meaningful contributions to new or ongoing at-risk butterfly recovery initiatives is critical and typically driven by available expertise and resources. A detailed decision tree is often a useful tool that can be employed to review available options.

References

Association of Zoos and Aquariums (2012) Species Survival Plan® (SSP) Program handbook. Association of Zoos and Aquariums, Silver Spring

Conservation Measures Partnership (2004) Open standards for the practice of conservation (Version 2.0). Washington, DC. http://cmp-openstandards.org/wp-content/uploads/2014/03/CMP-OS-V3-0-Final.pdf. Accessed 6 Jan 2015

Crouse DT, Mehrhoff LA, Parkin MJ, Elam DR, Chen LY (2002) Endangered species recovery and the SCB study: a U.S. Fish and Wildlife Service perspective. Ecol Appl 12:719–723

Daniels JC (2009) Cooperative conservation efforts to help recover and endangered south Florida butterfly. Insect Conserv Divers 2:62–64

Flockhart DTT, Pichancourt JB, Norris DR, Martin TG (2014) Unravelling the annual cycle in a migratory animal: breeding-season habitat loss drives population declines of monarch butterflies. J Anim Ecol. http://dx.doi.org/10.1111/1365-2656.12253

Fox R, Asher J, Brereton T, Roy D, Warren M (2006) The state of butterflies in Britain and Ireland. Pisces Publications, Oxford

Glowka L, Burhenne-Guilman F, Synge H, McNeely JA, Gündling L (1994) A guide to the convention on biological diversity. Environmental Policy and Law paper No. 30. IUCN. Gland, Switzerland

IUCN (2002) IUCN Technical Guidelines on the Management of Ex Situ Populations for Conservation. IUCN, Gland, Switzerland. https://portals.iucn.org/library/efiles/documents/Rep-2002-017.pdf. Accessed 17 Nov 2014

National Research Council (1992) Restoration of aquatic ecosystems: science, technology and public policy. National Academy Press, Washington, DC

NatureServe (2014) A network connecting science with conservation. http://www.natureserve.org. Accessed 2 Dec 2014

New TR (1991) Butterfly conservation. Oxford University Press, Oxford

Salafsky N, Margoluis R, Redford KH, Robinson JG (2002) Improving the practice of conservation: a conceptual framework and research agenda for conservation science. Conserv Biol 16(6):1469–1479

Schultz CB, Russel C, Wynn L (2008) Restoration, reintroduction, and captive propagation for at-risk butterflies: a review of British and American conservation efforts. Israel J Ecol Evol 54:41–61

Smith RK, Sutherland WJ (2014) Amphibian conservation: global evidence for the effects of interventions. Pelagic Publishing, Exeter

Soulé ME, Gilpin M, Conway W, Foose TJ (1986) The millennium ark: how long a voyage, how many staterooms, how many passengers? Zoo Biol 5:101–113

Species at Risk Public Registry. http://www.registrelep.gc.ca/species/schedules_e.cfm?id=1. Accessed 10 Nov 2014

Sutherland WJ, Armstrong D, Butchart S, Earnhardt JM, Ewen J, Jamieson I, Jones CG, Lee R, Newberry P, Nichols JD, Parker KA, Sarrazin F, Seddon PJ, Shah N, Tatayah V (2010) Standards for documenting and monitoring bird reintroduction projects. Conserv Lett 3:229–235

Synder NFR, Derrickson SR, Beissinger SR, Wiley JW, Smith TB, Toone WD, Miller B (1996) Limitations of captive breeding in endangered species recovery. Conserv Biol 10(2):338–348

Thomas JA, Clarke RT (2004) Extinction rates and butterflies. Science 305:1563–1564

Thomas JA, Telfer MG, Roy DB, Preston CD, Greenwood JJD, Asher J, Fox R, Clarke RT, Lawton JH (2004) Comparative losses of British butterflies, birds, and plants and the global extinction crisis. Science 303:1879–1881

Thomas JA, Simcox DJ, Clarke RT (2009) Successful conservation of a threatened Maculinea butterfly. Science 325:80–83

Tolson PJ, Magdich ML (2001) Building alliances for the conservation of oak savanna butterflies. In: Annual conference proceedings of the American Zoo and Aquarium Association, Saint Louis, 7–11 September 2001

U.S. Fish and Wild Service (2000) Policy regarding controlled propagation of species listed under the endangered species act. 65 FR 56916, September 20, 2000

US Fish and Wildlife Service (2014) Endangered and threatened wildlife and plants; endangered status for the Florida Leafwing and Bartram's scrub-hairstreak butterflies, final rule. Federal Register 79: 155 (August 12, 2014), pp 47222–47244

US Fish and Wildlife Service, Endangered and threatened wildlife and plants, U.S. Government Printing Office, 50 eCFR 17.11 and 17.12. http://www.ecfr.gov/cgi-bin/text-idx?c=ecfr&sid=186cb0f38a1b1b6770e432a7eba20553&rgn=div8&view=text&node=50:2.0.1.1.1.2.1.1&idno=50. Accessed 9 Dec 2014

van Swaay CAM, van Strien AJ, Harpke A, Fontaine B, Stefanescu C, Roy D, Maes D, Kühn E, Õunap E, Regan E, Švitra G, Prokofev I, Heliölä J, Settele J, Pettersson L, Botham L, Musche M, Titeux N, Cornish N, Leopold P, Julliard R, Verovnik R, Öberg S, Popov S, Collins S, Goloshchapova S, Roth T, Brereton T, Warren M (2013) The European grassland butterfly indicator: 1990–2011. European Environment Agency, Copenhagen, 34 pp

The Role of AZA-Accredited Zoos and Aquariums in Butterfly Conservation

Shelly Grow, Ruth Allard, and Debborah Luke

Abstract Zoos and aquariums accredited by the Association of Zoos and Aquariums (AZA) have long engaged in conservation with activities ranging from ex situ propagation of animals for reintroduction, to in situ field conservation, to developing conservation education programs promoting conservation actions linked to wildlife issues, to undertaking important scientific research. Furthermore, AZA-accredited zoos and aquariums are significant funders of conservation and, with support and engagement by the Association itself, their involvement will continue to grow. Butterflies in North America have been the beneficiaries of these efforts, supported in part by the creation of the Butterfly Conservation Initiative in 2001, a collaboration intended to grow the butterfly conservation community of practice among AZA-accredited zoos and aquariums and partner organizations. Between 2010 and 2013, AZA members spent more than $2 million on butterfly conservation, including major initiatives supporting ten vulnerable species as well as other conservation, education, and research efforts throughout North America. These financial and staff investments shows that AZA-accredited facilities have the experience, expertise, and motivation to make a difference on behalf of butterflies.

Introduction: The Role of Zoos and Aquariums in Conservation

Zoos and aquariums accredited by the Association of Zoos and Aquariums (AZA) serve as conservation centers that are concerned about ecosystem health, take responsibility for species survival, contribute to research, conservation, and education, and provide society the opportunity to develop personal connections with the animals in their care (Hutchins and Smith 2003). These popular organizations have

S. Grow (✉) • D. Luke
Association of Zoos and Aquariums, Silver Spring, MD, USA
e-mail: SGrow@aza.org; DLuke@aza.org

R. Allard
Phoenix Zoo/Arizona Center for Nature Conservation, Phoenix, AZ, USA
e-mail: RAllard@phoenixzoo.org

© Springer Science+Business Media B.V. 2015

J.C. Daniels (ed.), *Butterfly Conservation in North America*,
DOI 10.1007/978-94-017-9852-5_2

made an appreciable contribution to biodiversity conservation (Gusset and Dick 2010) while helping the public appreciate and participate in conservation (Clayton et al. 2009). AZA Accreditation Standards and Board-approved policies ensure that all accredited facilities engage in conservation, although participation varies by organization (Gusset and Dick 2010; Association of Zoos and Aquariums 2014a).

Zoos and aquariums have long been seen as contributors to species recovery (Barber 2008). Since the 1960s, iconic animals worldwide have been identified as endangered – or even extinct in the wild – and their ex situ conservation breeding and reintroduction over the decades has helped recover and support their populations in the wild (Conde et al. 2011). For example, the Przewalski's horse (*Equus ferus*) in northwestern China was identified as extinct in the wild while the golden lion tamarin (*Leontopithecus rosalia*) in coastal Brazil was identified as critically endangered but now, due in part to the efforts of AZA-accredited zoos, both have had their threat status reduced to endangered (Kierulff et al. 2008; Boyd and King 2011). The Arabian oryx (*Oryx leucoryx*) was identified as extinct in the wild in the 1972, but after decades of conservation work led in large part by AZA-accredited organizations, this species became the first to improve its population status by three full categories on the IUCN Red List and was re-classified as vulnerable in 2011 (IUCN 2011).

Species recovery is an endeavor that includes both global efforts and work focused domestically within the United States. It is thanks in part to the ex situ work of AZA-accredited zoos and aquariums that the United States can still proudly count black-footed ferrets (*Mustela nigripes*), California condors (*Gymnogyps californianus*), and many other animals among its wildlife ranks (Hutchins and Wiese 1996; Belant et al. 2008; BirdLife International 2013). However, due to the challenges of managing ex situ populations, financial costs, disease concerns, limited habitat availability, and other issues, conservation breeding and/or rearing is only appropriate for a subset of species facing declines (Hutchins and Conway 1995; Snyder et al. 1996). Ex situ programs can be a useful conservation tool, but only once extensive field evaluations have been conducted, as part of comprehensive programs to maintain or restore populations in wild habitats, and ideally before the species has declined to only a few remaining individuals (Snyder et al. 1996; Bowkett 2009; Conde et al. 2011).

AZA-accredited zoos and aquariums do much more than save species on the brink of extinction through ex situ breeding or rearing. They also coordinate, participate in, and support conservation programs around the world, provide veterinary and rehabilitative care in the face of wildlife disease and injury, conduct both lab and field research on ways to protect species, habitats, and ecosystems in the wild, assist wildlife agencies in surveying and monitoring wild populations, and work in range countries and beyond to increase conservation awareness, capacity, advocacy, and action. Furthermore, the global zoo and aquarium community is one of the main providers of conservation funding worldwide (Gusset and Dick 2011). Since 2010, the AZA has surveyed its members about conservation activities that directly impact animals and habitats in the wild. Responses are reviewed against a set of criteria developed by AZA's Field Conservation Committee (Association of Zoos and Aquariums 2013a) and have shown that AZA member organizations spend approximately $160 million annually on projects taking place in more than 100 countries (Association of Zoos and Aquariums 2014a).

The role of AZA-accredited zoos in aquariums in conservation extends beyond those activities having a direct impact on animals and habitats in the wild. Because resource conservation can lessen some of the most significant threats facing wildlife worldwide (e.g., habitat loss and degradation) many AZA-accredited zoos and aquariums are striving to conserve the natural resources utilized in their business operations (AZA Green Scientific Advisory Group 2013; Gonzalez Ciccia 2014). Sustainable (green) business practices are being implemented to reduce water and energy usage, decrease waste generation, and encourage the generation and purchase of green products and renewable energy (Grow et al. 2013). Capital investments are being made so that the daily operations of AZA-accredited zoos and aquariums are aligned with their conservation missions.

Furthermore, a commitment to scientific research, both basic and applied, is a trademark of the modern zoological park and aquarium (Hutchins and Smith 2003). AZA-accredited zoos and aquariums conduct or facilitate research investigations to mitigate human-wildlife conflict, improve reintroduction methodologies, develop new tools for monitoring animals in the wild, examine disease transmission between domestic and wild animals, test bird-friendly options for zoo and aquarium construction, and more (Association of Zoos and Aquariums 2013b). Research conducted at, or in coordination with, accredited zoos and aquariums in areas such as sensory biology, physiology, nutrition, reproduction, and cognition has increased understanding of the species being investigated and provides results that may be applied to benefit the health of animals in wild populations. In order for conservation activities to achieve their intended outcomes, AZA-accredited zoos and aquariums rely on research and science (Hutchins and Thompson 2008; Association of Zoos and Aquariums 2013b).

Finally, education and awareness-raising are essential elements of the culture of AZA-accredited zoos and aquariums (Hutchins and Conway 1995; Patrick et al. 2007), which collectively welcome 180 million guests annually. Animals cared for in AZA-accredited zoos and aquariums tend to be charismatic, less-threatened species (Fa et al. 2014), and exhibits are complemented by high quality education programs. Some of these programs deliver generalized messages, while other innovative conservation education programs both raise awareness about a conservation issue and promote conservation action (Association of Zoos and Aquariums 2014b). Given that the global zoological community is estimated to reach over 700 million visitors each year, there is great potential for these organizations to educate and inspire the public to engage in behaviors supporting conservation (Zimmerman et al. 2007; Fa et al. 2014).

Conservation is a priority for AZA-accredited institutions and is a key component of their missions. Whether saving species on the brink of extinction or ensuring species never reach such a precarious state, AZA-accredited zoos and aquariums are positioned well to protect wildlife and wild habitats for future generations to enjoy, and in fact consider this role an essential responsibility of any modern, professionally managed zoological facility (Hutchins and Smith 2003).

Why Butterfly Conservation Is a Good Fit for Zoological Organizations

Butterfly conservation has great appeal for AZA-accredited zoos and aquariums, for several reasons (Allard et al. 2001). Due to the fact that many butterfly conservation programs are local or regional in scope, they may entail a lower financial cost for participation. For those organizations managed by government agencies (e.g., municipal parks departments), expending funds to support local conservation and environmental priorities may be particularly logical (Hutchins and Conway 1995). Finally, engaging in butterfly conservation offers organizations, both with or without butterfly exhibits, opportunities to engage visitors and their communities (Sullivan 2003).

Development of BFCI

AZA-accredited zoos and aquariums, together with federal agencies, universities, and other non-profit organizations, formed the Butterfly Conservation Initiative (BFCI) in 2001 to capitalize on a convergence of ideas and enthusiasm. BFCI's founding objectives were to support federally listed butterfly species recovery and increase public awareness of and involvement in local and regional butterfly conservation efforts (Allard et al. 2001). It had long been recognized that the cooperative management and conservation of invertebrates was underrepresented in the zoological community (Cooper 1986; Hughes and Bennett 1991). Furthermore, while larger facilities were leading, funding, and/or otherwise deeply engaged in conservation efforts around the world, smaller and mid-sized facilities were seeking ways in which they could also contribute meaningfully to conservation on a scale consistent with their budget and staffing capacities (Allard and Hutchins 2003). Simultaneously, the larger AZA community, encompassing accredited zoos and aquariums of all sizes, recognized both the need and the opportunity to demonstrate their commitment to the conservation of wildlife in their own communities and not only in the international sphere (Allard and Hutchins 2003). The BFCI emerged from numerous meetings and discussions between AZA and the U.S. Fish and Wildlife Service's Office of Partnerships, while proof of concept activities and strategic planning further demonstrated the potential of this collaboration (Allard et al. 2001; Allard 2002).

Conceptual discussions focused on the butterfly species listed under the US Endangered Species Act, but the BFCI's mission was ultimately broadened to include "the conservation of threatened, endangered, and vulnerable butterflies and the habitats that sustain them, with a focus on research, recovery, and education"

(Allard and Hutchins 2003). This mission had sufficient resonance that BFCI quickly gained a membership of more than 50 AZA-accredited zoos and aquariums and six partner organizations, including the Association of Zoos and Aquariums (previously called the American Zoo and Aquarium Association), Environmental Defense, National Wildlife Federation, U.S. Fish and Wildlife Service, the Xerces Society, and the McGuire Center for Lepidoptera and Biodiversity at the University of Florida. The collaborative was financially supported by BFCI membership dues and early grants from the Disney Worldwide Conservation Fund (previously called the Disney Wildlife Conservation Fund), National Fish and Wildlife Foundation, U.S. Fish and Wildlife Service, and George Gund Foundation.

BFCI began implementation of activities dedicated to its mission immediately. Early pilot activities included bringing stakeholders together for a Karner Blue Butterfly (*Lycaeides melissa samuelis*) Recovery Implementation Workshop and developing a Butterfly Activity Guide for zoo and aquarium educators. Over the years, additional collaborative activities arose and evolved. The Butterfly Recovery Needs Matrix, originally published on compact disc to help partners identify those species in most urgent need of conservation attention, evolved into an online tool hosted by the Xerces Society (Walker 2003). Faculty at the McGuire Center for Lepidoptera and Biodiversity launched the statewide Florida Butterfly Monitoring Network with the engagement of several BFCI members and some financial support from AZA's Conservation Grants Fund (previously called AZA's Conservation Endowment Fund). "Techniques for Butterfly Conservation and Management," a training program developed and offered by the AZA's Professional Development Committee (previously called AZA's Board of Regents), served as a template for the much more in-depth, Institute of Museum and Library Services (IMLS)-funded, "Imperiled Butterfly Conservation and Management" course led by the McGuire Center for Lepidoptera and Biodiversity and including instructors and students from several BFCI member facilities. Additional activities included regional stakeholder meetings in the Pacific Northwest (Grow et al. 2004) and California, a small grants program, and communication vehicles to support a growing community of practice focused on issues such as butterfly gardening/habitat creation, temporary and permanent butterfly exhibits, education and outreach programs, tagging of migrating butterflies, and collaboration on ex situ and field conservation projects.

In 2006, the hosting organization and fiscal agent for BFCI transitioned from AZA to the McGuire Center for Lepidoptera and Biodiversity and the Florida Biodiversity Foundation. Since 2013, its fiscal agent has been the University of Florida Foundation.

Conservation in the Field

Between 2010 and 2013, 44 AZA-accredited zoos and aquariums reported spending more than $2 million on Lepidoptera-related conservation efforts (Association of Zoos and Aquariums 2014c). Species-specific efforts dedicated to ten species accounted for more than $1.4 million of this sum, while the remaining amount was spent on landscape-focused butterfly projects (e.g., prairie butterfly conservation by the Minnesota Zoo) or on butterfly surveys and monitoring efforts. Butterfly conservation activities took place almost exclusively in North America, although approximately $20,000 was spent on butterfly conservation activities in three countries in South America. The species that received the most financial support was the federally threatened Oregon silverspot (*Speyeria zerene hippolyta*) at almost $750,000 over these 4 years, followed by the federally endangered Taylor's checkerspot butterfly (*Euphydryas editha taylori*) at approximately $300,000. The federally endangered Karner blue butterfly benefitted from more than $200,000, plus a partnership that included seven AZA-accredited partners (including Binder Park Zoo, Buttonwood Park Zoo, Detroit Zoo, Museum of Science, Roger Williams Park Zoo, Toledo Zoo, and Toronto Zoo).

As the community of practice has grown, as well as the expertise of individual members and the trust they have built with local partners, so have the types of programs and engagement represented. For the Oregon Zoo, successful ex situ rearing and reintroduction of the Oregon silverspot butterfly led to partnerships with other BFCI members, including the Woodland Park Zoo and Oregon Coast Aquarium, as well as local agencies, universities, and both non-profit organizations and for-profit companies (Shepherdson and Csuti 2001; Grow et al. 2005). The Oregon silverspot collaboration received an AZA Significant Achievement in North American Conservation award in 2012 and, in that same year, the partnership grew to include the Pelican Pub & Brewery, which created Silverspot IPA to raise funds for the on-going conservation efforts (Sullivan 2013). What started with the ex situ propagation of a single species grew to include habitat restoration efforts and expanded to support the recovery of other species, including the federally endangered Fender's blue butterfly (*Icaricia icarioides fenderi*) and Taylor's checkerspot, and the Mardon skipper (*Polites mardon*), a candidate for listing under the Endangered Species Act (Grow et al. 2005).

Similarly, in the Great Lakes region, the Toledo Zoo had long been engaged in recovery efforts on behalf of the Karner blue butterfly, developing ex situ rearing protocols and releasing the butterfly into Ohio, where it had been extirpated (Tolson and Magdich 2001; U.S. Fish and Wildlife Service 2003). This program, which received AZA's Top Honors award for North American Conservation in 2003, remains vulnerable to stochastic weather events and reliant on active habitat management (Magdich 2008). Karner blue butterfly recovery efforts have engaged several other AZA-accredited facilities, some of which have engaged in releases and/or surveys (e.g., the Detroit Zoo and Binder Park Zoo) and others which have participated in host plant propagation and habitat restoration (e.g., Roger Williams Park Zoo and Buttonwood Park Zoo) (Association of Zoos and Aquariums 2014c). Roger Williams Park Zoo's engagement in Karner blue butterfly earned it AZA's Significant Achievement award for North American Conservation in 2010. The Zoo is not located within the butterfly's historic range, but has overseen the annual planting of lupine in a restoration area in New Hampshire, assisted with collection of lupine seed, propagated lupine, and reared Karner blue butterfly larvae for reintroduction.

In 2005, the Toledo Zoo expanded its butterfly conservation program to initiate a wetland butterfly initiative targeting three species; the purplish copper (*Lycaena helloides*) and the swamp metalmark (*Calephelis muticum*), which are both listed as endangered by the State of Ohio, and the federally listed Mitchell's satyr (*Neonymphya mitchelli*). While this initiative included a conservation rearing and release element, it also included the development of inquiry-based elementary-aged wetland butterfly classroom curricula aligned to Michigan academic content standards in science. Besides the Toledo Zoo, three additional AZA-accredited zoos in Michigan (the Detroit, John Ball, and Binder Park Zoos) engaged in this project by also joining the U.S. Fish and Wildlife Service's Mitchell's Satyr Working Group (Grow et al. 2005).

Between 2010 and 2013, other AZA-accredited zoos participated in or laid the groundwork for translocations, rearing, and/or conservation breeding efforts on behalf of additional butterflies and their host plants (Association of Zoos and Aquariums 2014c). These efforts include the Brevard Zoo and Zoo Miami rearing and/or translocating the Atala butterfly (*Eumaeus atala*) in Florida. Further north, ZOOAMERICA North American Wildlife Park has raised food plants and regal fritillary butterflies (*Speyeria idalia*) to supplement the single remaining population in Pennsylvania and re-establish other populations of the butterfly within the state, where it is listed as a species of concern. Biologists at the Minnesota Zoo have been studying wild populations of prairie butterflies in Minnesota, including the federally endangered Poweshiek skipperling (*Oarisma poweshiek*) and the federally threatened Dakota skipper (*Hesperia dacotae*), assessing their population status in the wild and the threats they face, in addition to creating assurance populations of the Dakota skipper (Minnesota Zoo 2014). To date, none of the butterfly conservation breeding or rearing programs are managed at the AZA level as a Species Survival Plan® (SSP) due to the regional scope of the programs and their resource intensive nature (AZA Terrestrial Invertebrate Taxon Advisory Group 2009).

While some facilities are engaged in conducting species-specific surveys, AZA-accredited zoos and aquariums have also participated in both local and statewide butterfly surveying and monitoring programs, including in Colorado (Cheyenne Mountain Zoo), Florida (Brevard Zoo, Disney's Animal Kingdom, Jacksonville Zoo and Gardens, Zoo Miami), Michigan (Detroit Zoo), Nebraska (Omaha's Henry Doorly Zoo), Ohio (Cleveland Metroparks Zoo), and Tennessee, (Knoxville Zoological Gardens) (Association of Zoos and Aquariums 2014c). Engagement has included helping to establish new monitoring networks, monitoring by staff, and recruiting and training volunteer citizen scientists (Grow and Allard 2006).

Several AZA-accredited zoos and aquariums have been actively engaged in several facets of monarch (*Danaus plexippus*) conservation. Disney's Animal Kingdom and the Disney Worldwide Conservation Fund, as well as SeaWorld San Diego have provided funding to support communities living within and around Central Mexico's Monarch Butterfly Biosphere Reserve. Other facilities have contributed by creating monarch habitat, supporting research, participating in monarch "counts", or engaging the public in the tagging of migrating monarchs. Support is also provided to the non-profit organization Monarch Watch to help recover butterfly tags each year. Facilities involved in these efforts between 2010 and 2013 include Cleveland Metroparks Zoo, Clyde Peeling's Reptiland, Disney's Animal Kingdom, Knoxville Zoological Gardens, Northeastern Wisconsin (NEW) Zoo & Adventure Park, Oglebay's Good Zoo, Oklahoma City Zoological Park, Phoenix Zoo, San Antonio Zoological Society, Santa Barbara Zoological Gardens, Topeka Zoo, and the Tulsa Zoo (Association of Zoos and Aquariums 2014c).

Education

AZA-accredited facilities are able to engage large swaths of the population in conservation action through innovative conservation education programs. In 2013, a survey of AZA-accredited zoos and aquariums indicated over 26 million instances of visitors participating in programming that presented a clear conservation action related to a wildlife concern (Association of Zoos and Aquariums 2014a). Citizen science monitoring and tagging programs have been one prominent way of engaging people and raising awareness about butterflies and their conservation needs. Pollinator gardens and plantings, both on-site and in local communities, have been another widely used tool both to create butterfly habitat and engage diverse audiences. Between 2010 and 2013, The Seas and the Disney Worldwide Conservation Fund provided annual support to the Xerces Society for creating pollinator habitat in agricultural landscapes, while the Chicago Zoological Society – Brookfield Zoo participated in local prairie plantings. The Seneca Park Zoo's "Butterfly Beltway" program launched in 2002, and by 2010 this on-going program had supported the creation or restoration of more than 8,000-square-feet of butterfly habitat at businesses, schools, daycare facilities, special needs services facilities, community centers, and senior residences in nine counties across western New York. The gardens serve both to inspire others to plant gardens and to reach more people with the message of butterfly conservation (Seneca Park Zoo 2010).

Butterfly exhibits, either permanent or seasonal, are offered at approximately 70 AZA-accredited zoos and aquariums and are another way to raise awareness about butterflies and their habitats (Butterfly Conservation Initiative 2014). Butterfly exhibits often include conservation messages, and may be accompanied by native plant garden workshops or planting recommendations. Like other immersive animal exhibits, these exhibits may be particularly well positioned to foster learning, as they offer the visitor interactive opportunities with live animals and plants and may appeal to a range of visitor motivations (National Research Council 2009). Evaluations conducted at the Woodland Park Zoo's seasonal exhibit, Butterflies and Blooms, demonstrated visitor enthusiasm, as well as alluded to the potential for engaging visitors in butterfly conservation. Visitors reported high satisfaction with the exhibit; over 40 % surveyed had visited the fee-based exhibit four or more times previously, and people ranked the exhibit third from the top on the list of favorites (Owen 2003). Over one-quarter of visitors responded to the prompts "Visiting the butterfly exhibit reminded me that . . . " and "I never realized that . . . " by mentioning their desire to spend more time in nature and wanting to attract butterflies to their own garden (Owen 2003).

Future Directions

Since the 1990s, AZA has supported an organizational infrastructure to facilitate collective field conservation (Hutchins and Conway 1995). This infrastructure includes an AZA Field Conservation Committee as well as SSPs and Taxon Advisory Groups (TAGs), which have been given responsibility for both animal management and conservation goals (Hutchins and Wiese 1991). However, engagement in field conservation still varies significantly by organization while conservation need has only grown. AZA and its Field Conservation Committee are committed to increasing meaningful engagement by all member organizations. AZA's Board of Directors passed a resolution in 2013 challenging each member to be seen individually as an organization committed to conservation of species in the wild, and is developing new initiatives intended to help members reach this goal (Association of Zoos and Aquariums 2013c). In 2014, one of the new initiatives, AZA SAFE: Saving Animals from Extinction, involved the creation of a conservation arm within the Association itself, directed specifically toward saving vulnerable species from extinction (Pate 2014).

Summary and Key Points

AZA-accredited zoos and aquariums have long been identified as conservation organizations, championing species and habitat protection and educating the public about how individual actions can impact the world around them. Their professional expertise and infrastructure have allowed AZA-accredited zoos and aquariums to serve as an "ark" for an array of animals and plants, breeding and rearing species of conservation concern in order to reintroduce, translocate, and supplement vulnerable populations. However, these facilities manage the animals in their care, education programs, and engagement in field conservation far beyond this single, albeit important, conservation tool. The animals cared for in AZA-accredited zoos and aquariums are vitally important to connect people to nature, provide education about conservation issues, inspire conservation action, and raise funds for species whose futures in the wild are increasingly uncertain. Science learning is facilitated through education programs and exhibits, while innovative conservation education programs link visitor learning to specific conservation issues and personal action. Beyond their ex situ role in conservation, AZA-accredited zoos and aquariums also provide significant conservation funding support, in addition to professional expertise on matters ranging from small population management, to creating or restoring habitat, to conducting scientific research and population status assessments, to engaging communities and building local capacity in order to reduce conservation threats. AZA-accredited zoos and aquariums are active and engaged conservation organizations that will continue striving to achieve a growing conservation impact into the future.

References

Allard RA (2002) The Butterfly Conservation Initiative: partnerships for recovery. In: 2002 Invertebrates in captivity conference proceedings, Tucson, 31 July–4 August 2002

Allard R, Hutchins M (2003) BFCI: not just another short-lived acronym. Communiqué, American Zoo and Aquarium Association, June:5–7

Allard R, Hutchins M, Harrelson D (2001) The Butterfly Conservation Initiative: partnerships for pollinators. In: Annual conference proceedings of the American Zoo and Aquarium Association, Saint Louis, 7–11 September 2001

Association of Zoos and Aquariums (2013a) Defining field conservation for the AZA community. https://www.aza.org/uploadedFiles/Conservation/The%20Definition%20of%20Conservation_FCC2011.pdf. Accessed 10 Nov 2014

Association of Zoos and Aquariums (2013b) Zoos & aquariums & you. Association of Zoos and Aquariums, Silver Spring, 36 pp

Association of Zoos and Aquariums (2013c) AZA board resolution. https://www.aza.org/uploadedFiles/Conservation/July2013BoardResolution.pdf. Accessed 24 Nov 2014

Association of Zoos and Aquariums (2014a) Annual report on conservation and science. https://www.aza.org/annual-report-on-conservation-and-science/. Accessed 10 Nov 2014

Association of Zoos and Aquariums (2014b) Defining education programs for the AZA community. https://www.aza.org/uploadedFiles/Conservation/Commitments_and_Impacts/DefiningEducationProgramsfortheAZACommunity-CEC2013.pdf. Accessed 10 Nov 2014

Association of Zoos and Aquariums (2014c) My AZA online conservation database. Silver Spring. https://www.aza.org/. Accessed 10 Nov 2014

AZA Green Scientific Advisory Group (2013) Green SAG update. Connect, Association of Zoos and Aquariums, May:30

AZA Terrestrial Invertebrate Taxon Advisory Group (2009) 2009–2011 TITAG regional collection plan. Association of Zoos and Aquariums, Silver Spring, 28 pp

Barber D (2008) Building conservation partnerships with zoos. Endanger Species Bull 33(1):18–19

Belant J, Gober P, Biggins D (2008) Mustela nigripes. In: The IUCN red list of threatened species, version 2014.2. http://www.iucnredlist.org. Accessed 13 Nov 2014

BirdLife International (2013) Gymnogyps californianus. In: The IUCN red list of threatened species, version 2014.2. http://www.iucnredlist.org. Accessed 13 Nov 2014

Bowkett A (2009) Recent captive-breeding proposals and the return of the ark concept to global species conservation. Conserv Biol 23(3):773–776

Boyd L, King SRB (2011) Equus ferus. In: The IUCN red list of threatened species, version 2014.2. http://www.iucnredlist.org. Accessed 13 Nov 2014

Butterfly Conservation Initiative (2014) Butterfly exhibits and gardens. http://butterflyrecovery.org/all-about-butterflies/butterfly-exhibits-and-gardens/. Accessed 12 Nov 2014

Clayton S, Fraser J, Saunders CD (2009) Zoo experiences: conversations, connections, and concern for animals. Zoo Biol 28:377–397

Conde D, Flesness N, Colchero F, Jones O, Scheuerlein A (2011) An emerging role of zoos to conserve biodiversity. Science 331:1390–1391

Cooper JE (1986) Captive breeding of invertebrates. Int Zoo Yearb 24/25:74–76

Fa J, Gusset M, Flesness N, Conde D (2014) Zoos have yet to unveil their full conservation potential. Anim Conserv 17:97–100

Gonzalez Ciccia P (2014) Sustainability in zoos and aquariums. Connect, Association of Zoos and Aquariums, February:18–19

Grow S, Allard R (2006) Magnifying education and conservation impact through citizen-science monitoring networks. In: 2006 invertebrates in captivity conference proceedings, Rio Rico, 26–30 July 2006

Grow S, Sullivan E, Allard R (2004) Working together and flying forward. Communiqué, American Zoo and Aquarium Association, April:38–39

Grow S, Tolson P, Shepherdson D, Allard R (2005) Passing through instars: butterfly conservation grows through experience. Communiqué, American Zoo and Aquarium Association, September:14–16

Grow S, Luke D, Hale J, Piekarz D (2013) How AZA in America is helping zoos and aquariums implement sustainable practices to meet their conservation missions. Zooquar Eur Assoc Zoos Aquaria 83:22

Gusset M, Dick G (2010) 'Building a future for wildlife?' Evaluating the contribution of the world zoo and aquarium community to in situ conservation. Int Zoo Yearb 44:183–191

Gusset M, Dick G (2011) The global reach of zoos and aquariums in visitor numbers and conservation expenditures. Zoo Biol 30:566–569

Hughes DG, Bennett PM (1991) Captive breeding and the conservation of invertebrates. Int Zoo Yearb 30:45–51

Hutchins M, Conway W (1995) Beyond Noah's Ark: the evolving role of modern zoological parks and aquariums in field conservation. Int Zoo Yearb 34:117–130

Hutchins M, Smith B (2003) Characteristics of a world-class zoo or aquarium in the 21st century. Int Zoo Yearb 38:130–141

Hutchins M, Thompson S (2008) Zoo and aquarium research: priority setting for the coming decades. Zoo Biol 27(6):488–497

Hutchins M, Wiese R (1991) Beyond genetic and demographic management: the future of the Species Survival Plan and related AAZPA conservation efforts. Zoo Biol 10:285–292

Hutchins M, Wiese RJ (1996) Black-footed ferret recovery program analysis and action plan. American Zoo and Aquarium Association, Bethesda

IUCN (2011) A grain of hope in the desert. http://www.iucn.org/knowledge/news/?7671/A-grain-of-hope-in-the-desert. Accessed 24 Nov 2014

Kierulff MCM, Rylands AB, de Oliveira MM (2008) Leontopithecus rosalia. The IUCN red list of threatened species, version 2014.2. http://www.iucnredlist.org. Accessed 13 Nov 2014

Magdich M (2008) Captive management and re-introduction of the Karner blue butterfly to the oak openings of northwest Ohio, USA. In: Soorae PS (ed) Global re-introduction perspectives: re-introduction case-studies from around the globe. IUCN/SSC Re-introduction Specialist Group, Abu Dhabi, viii + 284 pp

Minnesota Zoo (2014) Saving prairie butterflies. http://mnzoo.org/conservation/minnesota/saving-minnesotas-prairie-butterfly-heritage/. Accessed 12 Nov 2014

National Research Council (2009) Learning science in informal environments: people, places, and pursuits. Bell P, Lewenstein B, Shouse AW, Feder MA (eds) Committee on Learning Science in Informal Environments, Board on Science Education, Center for Education, Division of Behavioral and Social Sciences and Education. The National Academies Press, Washington

Owen K (2003) Butterflies & blooms: evaluating the impact of an immersion experience. In: Annual conference proceedings of the American Zoo and Aquarium Association, Columbus, 7–11 September 2003

Pate D (2014) Message from the chair of the board. Connect, Association of Zoos and Aquariums, December:3

Patrick P, Matthews C, Ayers D, Tunnicliffe S (2007) Conservation and education: prominent themes in zoo mission statements. J Environ Educ 38(3):53–60

Seneca Park Zoo (2010) Butterfly beltway project. http://senecaparkzoo.org/Content/upload/media/Butterfly_Beltway.pdf. Accessed 12 Nov 2014

Shepherdson DJ, Csuti B (2001) Oregon silverspot butterfly population supplementation. In: Annual conference proceedings of the American Zoo and Aquarium Association, Saint Louis, 7–11 September 2001

Snyder N, Derrickson S, Beissinger S, Wiley J, Smith T, Toone W, Miller B (1996) Limitations of captive breeding in endangered species recovery. Conserv Biol 10(2):338–348

Sullivan E (2003) Butterflies: conservation in living color. Communiqué, American Zoo and Aquarium Association, June:10–11

Sullivan E (2013) Brewing for butterflies. In: Lewthwaite T (ed) Zoos & aquariums & you. Association of Zoos and Aquariums, Silver Spring, p 15

Tolson PJ, Magdich ML (2001) Building alliances for the conservation of oak savanna butterflies In: Annual conference proceedings of the American Zoo and Aquarium Association, Saint Louis, 7–11 September 2001

U.S. Fish and Wildlife Service (2003) Final recovery plan for the Karner blue butterfly (Lycaeides melissa samuelis). U.S. Fish and Wildlife Service, Fort Snelling, 273 pp

Walker H (2003) BFCI partners: who they are and what they do. Communiqué, American Zoo and Aquarium Association, June:37–38

Zimmerman A, Hatchwell M, Dickie LA, West C (2007) Zoos in the 21st century: catalysts for conservation? Conservation biology (No. 15). Cambridge University, Cambridge Press, Cambridge

Butterfly Monitoring for Conservation

Doug Taron and Leslie Ries

Abstract Monitoring butterfly populations is an essential component of their conservation. Some survey techniques measure occupancy, and simply determine the presence or absence of species, whereas other techniques measure butterfly abundance. Mark release recapture techniques involve marking the wings of a subset of a population, releasing and then recapturing them, and determining the proportion of marked individuals in the re-sampling. Distance sampling takes advantage of the decrease in probability of detection of individual butterflies as a function of increased distance from the observer. These techniques can both be used to estimate actual population size. Mark release recapture is the most rigorous, but also the most labor-intensive technique. It also carries risk of damage to individuals during the marking process. Distance sampling is statistically robust and doesn't risk damaging butterflies by marking them. In some cases, the requirement for survey transects to be placed randomly within the population, and the assumption that the butterflies are distributed uniformly limit the application of the technique. For Pollard walks, surveyors walk a set route at a uniform pace. They count all butterflies within a prescribed distance (generally about 20 m). In addition to these systematic survey techniques, a variety of less formal monitoring protocols are also used. These include count circles, field trips, and wandering surveys. There are also a wide variety of online opportunities for interested individuals to submit butterfly observations. Researchers should consider the assumptions, advantages and disadvantages when selecting a technique.

D. Taron (✉)
The Chicago Academy of Sciences, Peggy Notebaert Nature Museum, Chicago, IL, USA
e-mail: dtaron@naturemuseum.org

L. Ries
Department of Biology, University of Maryland, College Park, Maryland and the National Socio-Environmental Synthesis Center, Annapolis, MD, USA
e-mail: lries@sesync.org

© Springer Science+Business Media B.V. 2015
J.C. Daniels (ed.), *Butterfly Conservation in North America*,
DOI 10.1007/978-94-017-9852-5_3

Introduction

Numerous facets of the conservation of imperiled butterflies require knowledge of population trends, which in turn requires knowledge of population size. Monitoring populations is a key aspect of management because it provides information about how populations are faring and may provide critical cues to when management should be modified (Holling 1978). Monitoring is usually motivated by one of two goals: (1) targeted monitoring of specific species or populations that are known to be in jeopardy or (2) general monitoring to track trends in the overall butterfly community. Here, we describe the different types of monitoring protocols which are most frequently used for both targeted and general butterfly monitoring. Targeted monitoring is usually carried out by agencies or organizations focused on the conservation of a single, often rare, species or a local population and several examples are described in detail in chapters "Developing a Rare Butterfly Database for Conservation Purposes: An Example in Florida Using Citizen Scientists" and "Managing Land for Butterflies" of this volume. Due to their popularity, monarch butterflies (*Danaus plexippus*) are also monitored, but those efforts have been exhaustively described elsewhere (Oberhauser et al. 2015; Ries and Oberhauser 2015).

General monitoring, where all species are monitored, can be carried out by agencies or academic programs but increasingly these types of programs are accomplished through networks of citizen science volunteers. These groups are organized at local, regional, national, or even continental scales. General butterfly monitoring has produced a wealth of ecological information about the dynamics of butterfly populations in Europe, where butterfly monitoring has been well-established for many years (van Swaay et al. 2008). We focus the majority of this chapter on the network of general monitoring programs that have been rapidly growing throughout North America and describe the efforts of a new group, The North American Butterfly Monitoring Network (www.nab-net.org), to support growth of monitoring efforts and provide resources to collect, manage and share data. Achieving these goals is vital because while butterfly monitoring in North America has grown, the use of the data have not. To date, the use of monitoring data, either for research or to guide direct conservation action, has been very slow, lagging behind both butterfly monitoring programs in Europe and various bird monitoring programs.

Occupancy vs. Abundance

There are two main classes of data collected by monitoring programs: abundance and occupancy data. As its name implies, abundance data are used to quantify the size or density of a particular population, whereas occupancy data simply determine presence or absence of a taxon on a particular site or in a particular cell of a survey grid. Any monitoring method that can measure abundance can also measure occupancy, but there are some monitoring methods that are only suitable to track occupancy dynamics or define ranges.

Abundance data are typically collected either by marking, releasing, and recapturing individuals in a study area, or by observing and counting individuals in a defined area or along a transect. More recently, there have been advances in statistics that allow the development of abundance indices from more casual protocols like field trips and counts (Isaac et al. 2014). In most respects, abundance information can constitute a more powerful data set than occupancy data because change in abundance is a more sensitive indicator of population dynamics (Royle et al. 2005). Depending on the details of the data collection method, abundance methods can determine demographics, mobility, and lifespan in ways that occupancy cannot.

Despite the greater potential for diverse analyses of abundance data, there are situations where occupancy has decided advantages. In particular, species present at very low densities or that are very difficult to detect may not be suitable for the collection of abundance data (Bried and Pellet 2012; MacKenzie et al. 2005). Furthermore, certain monitoring techniques are only able to capture occupancy.

The following sections will examine in more detail several approaches to the collection and use of both abundance and occupancy data. Assumptions, strengths and weaknesses, and comparisons of the use and efficacy of the various methods will be discussed. We separate the all the different types of monitoring programs into two groups: systematic surveys, which employ more rigorous protocols, and informal methods such as counts, field trips, and sightings programs.

Systematic Surveys

Systematic surveys are those with the strictest protocols. Survey sites are established and usually visited multiple times within and between seasons. The same monitor or group of monitors often return to the same site and perform the same roles. Finally, the area is usually searched more exhaustively and the exact area of search is usually known. For this reason, these systematic surveys are more similar to academic protocols, with the major exception that random placement of survey sites is rarely used. However, overall, the protocols are usually much more consistent between surveys and thus much more comparable, making patterns over space and time much easier to describe.

Pollard Walks

Overview

Pollard walks, sometimes called Pollard transects, are named for Ernest Pollard, who pioneered the technique (Pollard 1977). Pollard's goal was to develop a technique that could be used to detect long-term changes in butterfly populations, and that could make use of recorders who might not have formal training in

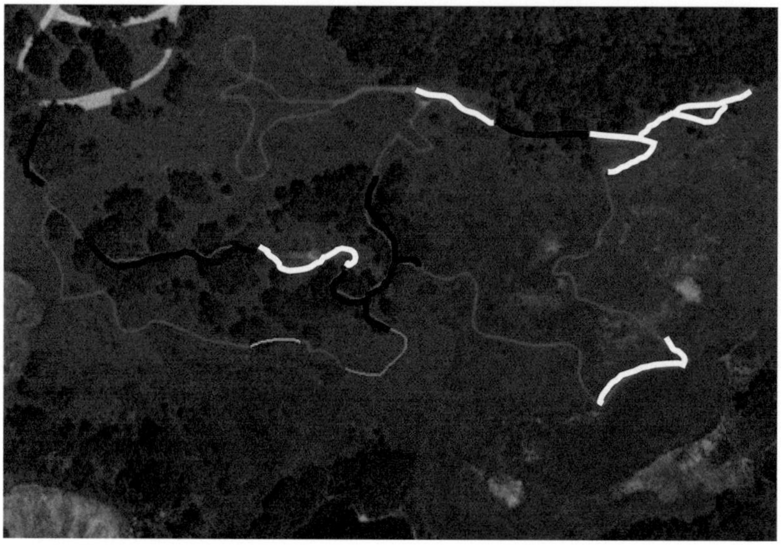

Fig. 1 Pollard routes are established as walking paths that often go through several habitat types. In the above example, the route is divided into several sections based on habitat type (shown by the different shades). For most programs, butterflies are recorded separately by habitat type (section), but those sections do not need to be contiguous. When surveys are performed, observers walk slowly along the route and observe butterflies only within a window of a set size, which may vary between programs

entomology. Pollard's techniques were developed at Monks Wood, a nature reserve in Cambridgeshire, England. Pollard employed his technique to evaluate effects of site management (Pollard 1982), phenology (Pollard 1991), and effects of weather (Pollard 1988) on butterflies, as well as to augment general understanding of butterfly ecology (Pollard and Yates 1993).

As initially described, the method involved walking a defined transect at weekly intervals and counting all butterflies seen at a defined distance (Fig. 1). The transects were divided into segments, and data were collected only under certain weather conditions (Pollard 1977). Since the initial description, many modifications to the protocol have been employed, typically involving the frequency of monitoring, acceptable weather conditions, and division of the walking route into subsections (Pollard and Yates 1993). Subsections are typically divided up by habitat (Fig. 1) although divisions are also made for other logistical reasons.

What It Measures

It's important to bear in mind that Pollard data do not return an actual population size, either over the entire generation of butterflies or restricted to the day of the survey. This is due to the fact that butterflies may be, and typically are, missed by the spotter (Haddad et al. 2008; Harker and Shreeve 2008; Pellet et al. 2012) or individuals may be counted multiple times. Although even a single Pollard survey will return information regarding distribution and phenology of butterflies, relative abundance data are most powerful when collected over extended period of time and used for detecting spatial or temporal changes in butterfly populations.

The Pollard method allows for correction for survey effort by adjusting by the length of the transect or by time, the latter requiring the recording of survey start and end times. This enables observations to be evaluated in terms of observations per unit time or distance. Because the technique is a general survey, it allows monitoring of most butterfly species.

Assumptions

An important assumption of the Pollard walk is that the counts are directly proportional to the actual population size (Haddad et al. 2008). Further, unless death rates are known, then integrating counts over time to develop a yearly index will not correctly account for population turnover within the season (Nowicki et al. 2008). Thus, yearly indices from uncorrected Pollard counts are often interpreted as the number of "butterfly-days" and not a true index of butterfly abundance, although new techniques to account for this have been proposed (e.g., Matechou et al. 2014). Despite these issues, there are various reports of differing degrees to which Pollard walk count indices do (Pollard 1977; Thomas 1983; Collier et al. 2008; Haddad et al. 2008; Isaac et al. 2011; Pellet et al. 2012), or do not (Harker and Shreeve 2008) correlate well with population size.

Advantages

One significant advantage to the Pollard approach to butterfly monitoring is its simplicity. The protocol, while rigorous, is uncomplicated, and can be readily taught to people who have little or no formal science training. The main challenge for nonscientists in conducting Pollard surveys is typically one of species identification rather than survey protocol.

Because this technique can be used by citizen scientists, the Pollard method has formed the basis of numerous citizen science programs that survey butterflies. The oldest is the British Butterfly Monitoring Scheme (Pollard and Yates 1993), founded in 1976 and now merged with other programs to be part of the United Kingdom Butterfly Monitoring Scheme (UKBMS) (www.ukbms.org). Since that time, the UKBMS has conducted more than 250,000 surveys and the program has expanded to more than ten European countries (van Swaay et al. 2008). The Illinois Butterfly Monitoring Network (Taron 1996), which also uses a protocol based on the Pollard walk, was founded in 1987. Through these efforts, a considerable body of literature has been amassed based on analyses of Pollard transect data. (See UKBMS website, www.ukbms.org/reportsAndPublications.aspx, for an extensive bibliography).

Limitations

Whereas the Pollard survey method is a robust technique that has produced a large body of data, multiple analyses, and numerous publications, it has several limitations and is not ideal in all situations. As previously mentioned, this survey technique is used to measure relative abundance rather than absolute population size.

Various deviations from the assumptions inherent to the technique have been reported. Many involve variations in detectability. Species that reside in the canopy of wooded areas, species with cryptic color patterns, and species such as some metalmarks that spend large amounts of time perched on the undersides of leaves will be detected with less efficiency than species with more conspicuous flight patterns (Shuey and Szymanski 2010; Isaac et al. 2011). Moreover, detectability of butterflies may vary in space, in time, or by species (Dennis et al. 2006; Gross et al. 2007; Harker and Shreeve 2008; Haddad et al. 2008).

Additionally, monitoring data are frequently applied to the detection of changes in butterfly population sizes that result from environmental changes such as those due to climate change or site management. Although these changes may well influence expansion or contraction of butterfly populations, they also affect vegetation in ways that may either enhance or hinder detectability, raising the possibility of confounding environmental and detectability effects (Isaac et al. 2011).

As is the case with many of the abundance techniques, species that are encountered on Pollard walks at very low densities may not always be detected as readily as more abundant species. Further, Pollard walks were designed to capture the maximum amount of information on the most species, whereas researchers wishing to monitor species that, for any of these reasons, are poorly detected by this method may adopt methods that are more tailored for those individual species. For instance, some species are best detected through surveys of leaf damage (e.g., the Karner blue butterfly). For species that locally exist in very low numbers researchers might wish to consider occupancy techniques as an alternative to abundance techniques.

Ease of Use/Time Commitment

A single Pollard transect can typically be run in anywhere between about an hour and about 3 h. In the various Pollard-based networks, individual surveys are conducted anywhere from a minimum of six times per season (Taron 1996) to weekly throughout the growing season (Pollard and Yates 1993; van Swaay 2008).

The larger data sets that are needed, for example, to detect long-term population trends have typically been collected by citizen science networks. Although the collection of Pollard transect data on a single site is a modest endeavor, developing and running a citizen science network will take considerably more planning and effort.

Timed Surveys

Timed surveys are similar in technique to Pollard walks and have many of the same assumptions, advantages and limitations. For most Pollard surveys, a transect is established and the amount of time it takes to complete the survey is typically recorded, but not set. In timed surveys, a survey area is usually defined and a "zig-zag" method for completely covering the area. Survey time is pre-set by the researcher and always the same between surveys within a particular project. This method can be ideal to better survey a more restricted area and capture a higher proportion of species and individuals (Kadlec et al. 2012).

Mark Release Recapture

Overview

In contrast to the Pollard method, which measures relative abundance of butterflies, mark release recapture (MRR) provides an estimate of actual population size. In this technique, a subset of members of a population is captured and marked, typically by writing on the wings with a fine-tip indelible marker. The marked individuals are released back into the population and a brief period of time is allowed to elapse so that the marked individuals can mix in with the remainder of the population. The population is then re-sampled and the numbers of marked and unmarked individuals in the sample are recorded. The proportion of marked individuals in the sample should be the same as the proportion of marked individuals in the population. Because the total number of marked individuals is known, this proportion can then be used to calculate the total population size.

Assumptions

Chief among the numerous assumptions in the MRR method is the requirement for a closed population – that during the survey, the population size does not change due to emergence of new adults, death, immigration, or emigration. Methods exist to correct for some of these effects (Pradel et al. 1997; Harker and Shreeve 2008). The method also assumes that there are no behavioral or detectability differences between marked and unmarked individuals.

What It Measures

MRR data are considered the best for obtaining true population estimates. In addition to measuring the size of a population, MRR can be used to study dispersal of marked individuals by observing where marked individuals are recaptured, making it a useful technique for studying dynamics of butterfly metapopulations (Hanski et al. 2000; Ricketts 2001; Polic et al. 2014).

Advantages

MRR has been described as "the most rigorous approach to population estimation because it incorporates the greatest amount of information into well-developed statistical methods" (Haddad et al. 2008). By providing a larger quantity of demographic information than transect counts, MRR allows for the possibility of estimating longevity, dispersal, and detectability of the taxa surveyed. Because death rates are known unlike for Pollard methods, within-season turnover can be accounted for when developing population estimates.

Limitations

In addition to those mentioned previously, limitations of MRR include the possibility of damaging individuals during the process of capture and marking (Murphy 1987). This limitation is of particular concern when working with small, vulnerable populations of species that are of conservation concern.

It is also possible that the process of capturing and marking individuals changes their propensity for recapture, either by changing their detectability or their susceptibility to recapture (New 1991; Haddad et al. 2008) relative to unmarked individuals. One MRR study of *Neonympha mitchellii francisci* showed no negative effects of handling on behavior or survival of marked butterflies (Kuefler et al. 2008), however other MRR studies have shown increased mortality (Morton 1982) and emigration (Singer and Wedlake 1981) following marking.

It is difficult to apply MRR to small species, such as those in the families Lycaenidae, Riodinidae, and Hesperiidae, because they can be very difficult to mark and are easily damaged. For small and low density populations it can be difficult to find and mark a sufficient number of individuals to allow marking a sufficient sample size.

Ease of Use, Time Commitment

MRR is a time consuming, labor-intensive and therefore relatively expensive method (Haddad et al. 2008). It requires training in correct handling and marking of the butterflies. For this reason, it has never been and is unlikely to ever be a protocol method adopted by a citizen science network and therefore large-scale or long-term data are almost never collected (Nowicki et al. 2008).

Distance Sampling

Distance sampling is a method for estimating population density which, when combined with knowledge of area occupied, can be used to calculate population size rather than relative abundance. The principle of distance monitoring is based on the decrease in detectability of a study organism as a function of distance from the observer. By fitting a curve to a histogram of observed individuals at increasing distance from the observer and integrating under the curve, density of individuals over the study area can be calculated, thus determining the total population size (Brown and Boyce 1998; Thomas et al. 2010; Isaac et al. 2011).

Assumptions

Distance sampling assumes that transects are placed randomly relative to butterflies (Haddad et al. 2008; Isaac et al. 2011), that the study organism is distributed uniformly along the transects from a distance of zero out to the limit of detectability (Haddad et al. 2008), that there is complete efficiency in detecting the organism at zero distance from the observer, that the study objects do not move, and that the distance measurements are exact (Thomas et al. 2010).

Advantages

Distance sampling estimates density rather than relative abundance, as does MRR, but without risking potential damage to sensitive species due to handling and marking them. Provided that the assumptions are met, it provides an unbiased estimate of population density (Brown and Boyce 1998; Thomas et al. 2010; Isaac et al. 2011). Rigorous in both protocol and statistics, distance sampling has an extensive track record of application in peer-reviewed studies (Isaac et al. 2011).

Limitations

Note that distance sampling improves estimates obtained within a single-transect, but does not account for the problem of within-season turnover, confounding yearly indices (see Pollard section above). In addition, many of the assumptions inherent in distance sampling limit its application. For example, due to site fragility, physical barriers to access, or non-uniformity of habitat it may not be possible to set transects that are randomly located within a population (Haddad et al. 2008). A threshold sample size of 60 required for accurate modeling limiting the utility of distance sampling in populations that are sparse or have low detectability (Thomas et al. 2010; Isaac et al. 2011; Pellet et al. 2012).

Ease of Use, Time Commitment

Transects for distance sampling are somewhat more involved to set up than are those for Pollard Walks, as they must be placed randomly in areas known to have uniform density of the study species. They provide a snapshot of population density at a particular time, and must be combined with another method, such as MRR, in order to determine population size across an entire generation (Haddad et al. 2008). Data collectors must be trained to collect accurate distance measurements (Thomas et al. 2010). Distance sampling is typically applied to a limited number of species in any given study due to the difficulty of collecting distance data on multiple species simultaneously (Isaac et al. 2011). Although distance sampling used to be the primary method of accounting for detectability, more modern techniques allow detectability to be estimated through occupancy modeling (Royle et al. 2005), although because of the change in abundance during the flight season, it may be difficult to meet some of the necessary assumptions of occupancy modeling.

Comparative Studies of Systematic Survey Methods

Because of the limitations of MRR, both in terms of the intensity of effort required and the potential for damaging individuals in the marking process, it would be desirable to be able to use transect-based counts as an alternative. Isaac et al. (2011) showed strong correlation between data obtained via distance sampling with estimates of absolute population size obtained with MRR. Other studies (Thomas 1983; Collier et al. 2008) reported similar findings in comparing distance sampling with MRR, however Harker and Shreeve (2008) and Pellet et al. (2012) reported poor correlation. Haddad et al. (2008) also compared yearly population indices with MRR and transect methods, finding such indices to have the advantage of producing survivorship estimates without the risks that accompany handling the butterflies for marking. The index methods produced higher variation than either MRR or Pollard, and may be of limited use for small populations. Several authors (Gross et al. 2007; Haddad et al. 2008; Pellet et al. 2012) have suggested that combining one of the transect sampling methods with a limited MRR study might allow for improved estimates while moderating the limitations of the MRR method.

Surveys That Use Informal Protocols

In contrast to the rigorous protocols used for systematic surveys, a number of methods use more informal protocols. These efforts are almost entirely restricted to citizen science volunteer programs. Fewer sampling parameters are defined, and protocols may permit more variability in the precise location of sampling, frequency of sampling, and number of observers in the survey team. In addition, observation areas are usually large relative to the actual area that is searched. These protocols are often used primarily as outreach to engage people in butterfly watching, but a complete inventory of all species observed and the number of each species observed are typically reported.

Count Circles

The Count Circle technique is employed by the North American Butterfly Association's (NABA's) Count Program. The program was started by the Xerces Society in 1975, but taken over by NABA in 1992. The protocol was adapted directly from the Christmas Bird Counts (Swengel 1990). Count circles are study areas 7.5 miles in radius and are surveyed by one to several parties (a party can be comprised of one to several individuals) within a single 24-h period, usually in June or July. Since 2008, monitors have been encouraged to conduct counts once in the spring, summer and fall. For that reason, the program is now officially known as the Seasonal Counts although the original name was the 4th of July Butterfly Counts (www.naba.org). Effort is measured by counting "party-hours" the sum of the total number of hours each party (regardless of individual numbers of observers) completes during the count day (Swengel 1990). Currently, counts must be conducted for a minimum of six party-hours by a minimum of four individuals, although before 2009, there was no minimum amount of participation.

The main advantages to the count circle method are the ability to generate very large data sets at large scales and a long history of data collection. Data have been recorded annually since 1975 (Swengel 1990) and through 2014 have produced data from over 10,000 surveys. Currently, there are about 400–450 count circles are conducted every year.

Advantages

The simple protocol and wide geographic scope of the program allow for the collection of large quantities of data. There is greater latitude available for the placement of count circle centers than there is in the placement of Pollard or distance sampling transects, allowing for surveys in locations that are less amenable to either of those two methods. Although this method uses less standardization than other methods, the data are amenable to detection of trends (O'Brien et al. 2011; Link et al. 2006).

Limitations

The simple protocol collects less information than other protocols including Pollard walks, which provide finer spatial resolution of observations. Most published studies have evaluated species that are conspicuous, widespread, and generally abundant, such as painted ladies (Vandenbosch 2003), monarchs (Swengel 1995; Koenig 2006), and a few other abundant species (Ries and Mullen 2008, however, see O'Brien et al. 2011). Most counts are only done once a year, although sometimes once each season (spring, summer and fall). Therefore, phenology (the pattern of flight across the season) has the potential to bias the development of abundance indices to a much greater degree than protocols that collect data multiple times within and across seasons.

Ease of Use

The program has been designed to be easy to use. Protocols, data entry and instructions for new participants to become involved in the program are well-established. The ease of use of the count circle method is essential for the method to express its main strength: the involvement of large numbers of participants in order to collect a very large data set.

Field Trips or Wandering Surveys

Field trips are also usually planned as an outreach or participation exercise, such as for a butterfly club. Trips are usually to designated locations like parks, and so the area of search is specified, but the specific areas that are covered during the field trip are often unknown. However, all species and individuals seen during the field trip are generally reported and time and number of observers are often also known. Although less rigorous than counts, new statistical techniques have been developed to compare population indices across years (Isaac et al. 2014). These types of data have been successfully used to track long-term trends (Bried and Pellet 2012).

Wandering Surveys have been used as a low-cost method of intensive survey of species of interest throughout their flight periods. They follow no fixed route, rather they attempt complete coverage of a site or of a particular habitat within a site (Longcore et al. 2010), typically with emphasis on detecting individuals of a single species or small suite of species that are of conservation concern. The survey designs are opportunistic, and survey routes frequently cannot be closely replicated from survey to survey. Wandering survey data of the mission blue and callippe fritillary in the San Bruno Mountains of California were useful when treated as occupancy rather than abundance data (Longcore et al. 2010).

Although wandering survey design "violates most tenets of survey design" (Longcore et al. 2010), such data can, nonetheless, be informative for butterfly conservation. Particularly when species of interest are present at very low density (Bried and Pellet 2012), or occupy large numbers of fairly discrete patches (WallisDeVries 2004), occupancy data may be a viable alternative to abundance data.

Sightings

Sightings refer to opportunistic reports of one or multiple species that are not part of a formal or even casual field trip, count, wandering survey or systematic survey. One of the distinguishing features of a sightings report is that nothing about effort is known; not the area of search, the time of search, or whether all observed butterflies seen were reported. For systems that only allow one sighting per record, it is assumed that not all individuals seen were reported. Several online databases currently exist that can accept this type of opportunistic data. Most current platforms require photographs (e.g., butterfliesandmoths.org), but not all (e.g., NABA's Butterflies I've Seen platform). Platforms can be specific to butterflies or general platforms like inaturalist.org, projectnoah.org, or observation.org. eButterfly (e-butterfly.org) is another sightings platform, but it also accepts a range of protocols, from trips to Pollard walks. The submissions to these platforms are generally vetted by experts for quality control. When photographs are supplied, identification is vetted; otherwise, only unlikely (out-of-season, out-of-range) reports are flagged. Sightings data are most useful to draw range maps (e.g., via niche modeling), although they may also be useful to track some phenological pattern, or to contribute to species inventories (Matteson et al. 2012).

Inventory of North American Butterfly Monitoring Programs

The oldest formal butterfly monitoring program in North America was started by Art Shapiro (Forister et al. 2011). This academic-based program has been collecting data on 11 transects across an elevational gradient in central California since 1972. Transects are surveyed every other week, always by Art Shapiro and, in most cases, only presence/absence data are collected. The NABA program described above remains the longest-running, largest-scale butterfly citizen science program in existence. In 1986, the Illinois butterfly monitoring program was launched, patterned after the UK Butterfly Monitoring Scheme. This program was the impetus for many other state and local based butterfly monitoring programs, including Ohio (started in 1995) and Florida (started in 2003). Since the mid-2000s, there has been a sharp increase in the number and types of general butterfly monitoring programs and resulting in approximately 1,000,000 records collected through monitoring programs by January 2015 (Fig. 2). These programs span the range of protocols described above and have been implemented at a range of geographic scales, from very local (e.g., ranches managed for conservation) to continental.

In this section, we provide an inventory of general butterfly monitoring surveys in North America (Table 1), with the goal to give as complete a list as possible of

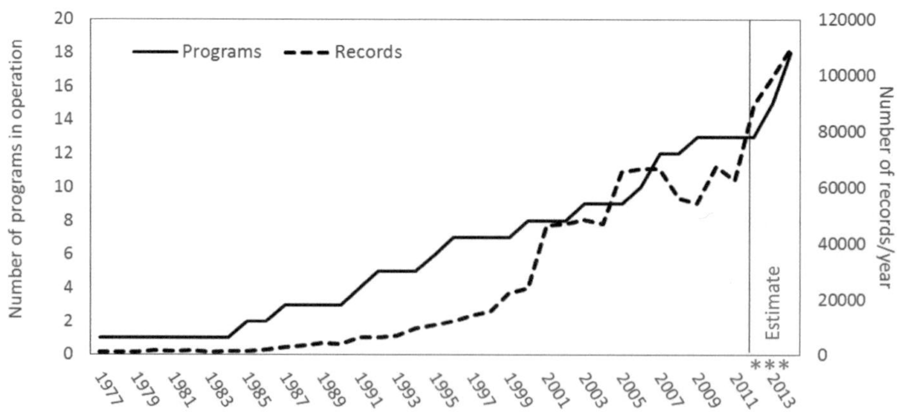

Fig. 2 An inventory of North American monitoring programs that survey the entire butterfly community (as opposed to a single species). Only citizen science programs are included here. Best estimates are given for 2012–2014 since not all records have been received from the programs. Just over 1,000,000 species records are predicted to be recorded through December 2014

Table 1 Known general butterfly monitoring programs in North America as of January 2015

Program (Acronym, start year)	Protocol type[1]	Program web address
(a) General butterfly programs: continental		
Butterflies and Moths of North America (BAMONA, 2005)	Sightings	www.butterfliesandmoths.org
eButterfly (eBFLY, 2011)	Trips/sightings	www.e-butterfly.org
North American Butterfly Association (NABA)		
Seasonal Count Program (1975)	Counts	www.naba.org/butter_counts.html
Butterflies I've Seen Program (2000)	Trips/sightings	www.nababis.org/
Recent Sightings Program	Sightings	http://sightings.naba.org/
(b) General butterfly programs: regional		
Carolina Butterfly Society (CA-LEPS, 2000)	Field trips	www.carolinabutterflysociety.org/
Cascades National Park BMN (C-BMN, 2011)	Pollard surveys	www.butterfliesandmoths.org/project/CNP
Colorado-BMN (CO-BMN, 2013)	Pollard surveys	No web site
Florida-BMN (FL-BMN, 2003)	Pollard surveys	www.flbutterflies.net/
Iowa-BMN (IA-BMN, 2007)	Pollard surveys	www.reimangardens.com/collections/insects/iowa-butterfly-survey-network/
Illinois-BMN (IL-BMN, 1987)	Pollard surveys	www.bfly.org
Massachusetts Butterfly Club (MBC, 1992)	Field trips	www.naba.org/chapters/nabambc/field-trips/asp
Michigan-BMN (MI-BMN, 2011)	Pollard surveys	www.michiganbutterfly.org/
Missouri-BMN (MO-BMN, 2014)	Pollard surveys	No web site
Ohio-BMN (OH-BMN, 1995)	Pollard surveys	www.ohiolepidopterists.org/bflymonitoring/
Tennessee-BMN (TN-BMN, 2014)	Pollard surveys	No web site
Swengel Monitoring (Swengel, 1985)	Timed surveys	No web site
(c) General butterfly programs: local		
Art Shapiro Monitoring (Shapiro, 1972)	Academic	http://butterfly.ucdavis.edu/
Boulder Open Space (BOS, 2007)	Surveys	No web site
Ft. Collins-BMN (FtC-BMN, 2014)	Surveys	No web site
Great Basin Monitoring (GBM, 1995)	Academic	No web site

(continued)

Table 1 (continued)

Program (Acronym, start year)	Protocol type[1]	Program web address
Greater Yellowstone Monitoring (GYM)	Academic	No web site
Irvine Ranch Conservancy-BMN (IRC-BMN, 2012)	Pollard surveys	No web site
MPG Ranch (MPG, 2014)	Pollard/timed	No web site
Occoquan Monitoring Program (OCC, 1991)	Pollard surveys	No web site
Rocky Mountain BMN (RM-BMN, 1995–2011)	Pollard surveys	http://www.nps.gov/romo/naturescience/rocky-mountain-butterfly-project.htm

Programs are separated into continental (a) and regional (b) and local (c) programs. Each record includes the program name, acronym (when referred to in other figures in this chapter), program type, start year, end year (if applicable), and program web address (if any). *BMN* Butterfly Monitoring Network. Monarch and other single-species focused programs are not included. Monarch-centric programs are described in detail in Oberhauser et al. 2015 and Ries and Oberhauser 2015.

[1]Note that this list only includes programs focused specifically on butterflies; general platforms like iNaturalist, Project Noah, and Observado are not included.

all programs currently in existence and to present several metrics of both the growth and activity of the programs. Our goal is to illustrate the scope of this growing resource so it can be better utilized by the conservation community. We focus here solely on general butterfly surveys, not those focused on individual species.

To understand the type of programs currently operating in North America, we arranged the programs on a diagram organized across two axis: strictness of protocols on the X-axis and geographic scope on the Y-axis (Fig. 3). Another interesting axis is time; however the majority of programs endeavor to operate in perpetuity so that metric becomes a function of when they were launched (indicated in Table 1). Note that the majority of programs cluster in two areas of the panel: large-scale programs with more casual protocols in the upper left-hand corner and regional and local programs following strict Pollard-like protocols in the mid- to lower-right hand corner. This clustering of programs makes sense as larger-scale programs are unlikely to be able to provide the training and volunteer management necessary for Pollard-based programs, which tend to be smaller-scale.

The North American Butterfly Monitoring Network

In May 2012, a meeting at the Socio-environmental Synthesis Center (sesync.org) in Annapolis, Maryland brought together leads of many long-lived, new, or developing butterfly monitoring programs. The group decided to form a network and thus was launched the North American Butterfly Monitoring Network (www.nab-net.org). The network has six goals which we review below. All of these goals are meant to expand participation, ease management of, and maximize the usefulness of butterfly monitoring data.

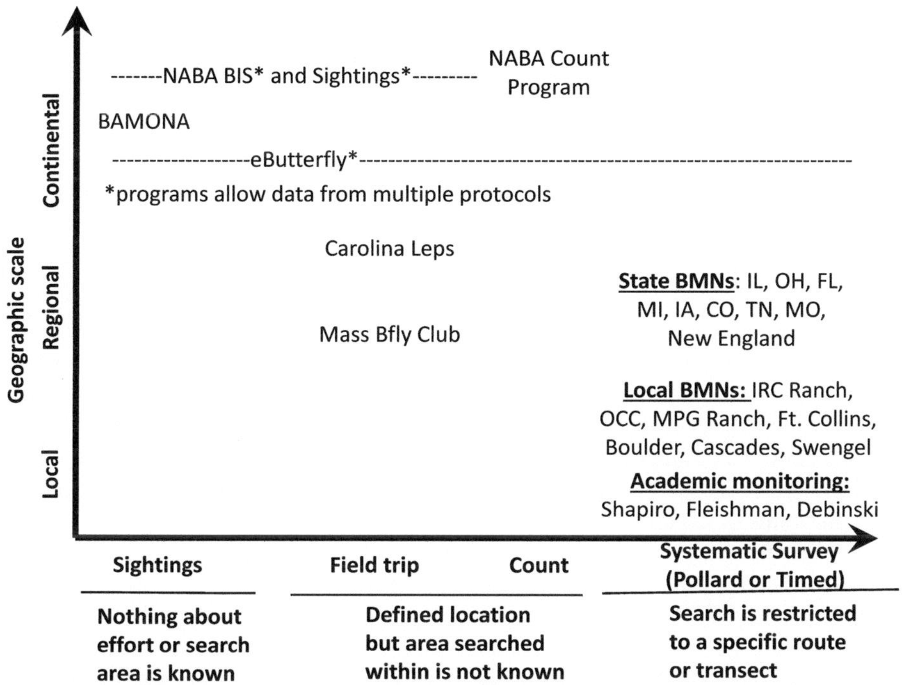

Fig. 3 Diagram showing all currently operating (as of January 2015) general butterfly survey programs organized by type of protocol (x-axis) and geographic scope (y-axis). eButterfly and NABA programs span multiple protocols and that is reflected in the lines (See Table 1. for start year and website information)

Goal 1: Track All North American Butterfly Monitoring

Keeping track of all programs collecting monitoring data on butterflies allows the public to find programs that they may be interested in volunteering for and makes it possible to discover butterfly data resources for scientists, those involved in conservation or policy, and the general public. The inventory as of January 2015 is shown in Table 1 and an updated list is curated at www.nab-net.org.

Goal 2: Standardize Protocols as Much as Possible

No program is compelled to adhere to specific protocol guidelines in order to be a member of nab-net. However, programs are encouraged to develop and describe their protocols, keep track of protocol changes, and tweak their protocols when possible to conform to best practices. Some protocol changes will affect data comparability pre- and post-change (e.g., changing the size of the detection window) while others will not (e.g., encouraging programs to start surveys earlier, end later, or perform more surveys during a year). When there is a long history of data collection, protocol changes that affect data comparability are discouraged.

Goal 3: Develop or Enhance Data Management Systems

All continental-scale programs (Fig. 3, Table 1) have online portals and systems to manage their data. However, the regional programs have a much greater challenge in collecting, managing, and sharing data especially since most are either run out of small institutions, often with small budgets, or are not affiliated with any institution. Further, the nature of the data collected through systematic surveys such as Pollard walks did not fit well on the major platforms, which would also not allow program managers to maintain control of the quality or use of their data. Indeed, data management was one of the biggest barriers to program growth, both for creating new programs or expanding existing ones.

In order to bridge this gap in a way that would benefit the maximum number of programs, the regional monitoring groups decided to collaborate with the group that runs Butterflies and Moths of North America to develop a platform specifically for regional programs that run Pollard-based surveys. The new platform, PollardBase, was launched in 2014 with seven of the programs using the system and most other programs planning to adopt it in the future. PollardBase has relieved much of the data management burden that most programs have by allowing volunteers to directly enter and proof their data, which are then vetted by program managers. Data can then be easily shared with local landowners, institutions, or the public. While the PollardBase initiative is working to meet the most critical need identified at the start of the network, there remains a broad goal to help support the best data management practices and make sure that tools are available and shareable between programs so that data collected by volunteers will not go to waste.

Goal 4: Share Data

Nab-net promotes the sharing of data and developing systems that promote sharing of data. Most programs are happy to share their data, but have challenges both in being able to format data (especially historical data for some programs) and also in developing robust data sharing policies. As of January 2015, few programs had yet been contacted to share their data, but as knowledge of butterfly monitoring programs has grown, requests for data were starting to come in. As of this date, the continental program whose data had been most requested for use in scientific publications on monarchs were NABA's and for regional programs were Illinois and Ohio's (Ries and Oberhauser 2015) and this is true for other species as well.

Goal 5: Expand Program Participation

Nab-net is developing or planning to develop a series of tools to promote program participation. To encourage new programs to start, there will be templates for timelines from program development, protocol guidance, training materials and access to systems like PollardBase. To expand participation of existing programs, there are guidelines to working with volunteers and land owners, all critical elements of running a sustainable program. In addition, by being part of a network, programs can inform their volunteers of other programs they may want to participate in.

Goal 6: Develop Analytical Tools

There have been many statistical models developed to analyze butterfly monitoring data, especially in Europe. However, many of these methods have difficulty being transferred to North American protocols because in general, most programs don't have as high a density of participants per unit area (so surveys are sparser on the ground) and volunteers also tend not to survey as frequently. That makes transferring modeling frameworks difficult. For instance, the most widely used approach to develop abundance indices for Pollard Surveys originally required transects to be visited nearly weekly (Rothery and Roy 2001), but then were adapted to allow as many as 30 % of weeks to be missed (Dennis et al. 2013). However, even that exceeds the visit rate of most North American surveys, so efforts are currently underway to adapt those protocols to allow even fewer visits per site by combining data across regions. This means that while site-specific indices may not be possible, regional ones will become the goal. Similar efforts to adapt methods for field trips and other opportunistic data are also underway. There are many statisticians and quantitative ecologists who are members of Nab-net and their goals are to take state-of-the-art methods and adapt them to North American Protocols.

Another effort at this time is to determine if death rates can be accounted for during transect counts. A method proposed by Zonneveld (1991) claimed to be able to do this and was even implemented in an online tool, INCA, which has many attractive features (Longcore et al. 2003). However, it turned out that death rate and population size were inseparable using those methods (Calabrese 2012). New methods have been proposed (Matechou et al. 2014) and are being tested using North American data.

Finally, a major goal of nab-net is to continue to track developments in statistical and analytical methods. As these methods evolve, members of nab-net will continually be seeking to adapt them to protocols in North America and make tools available that will make allow these techniques to be used by a variety of stakeholders.

Summary and Key Points

- Monitoring the sizes of butterfly populations is an essential component of conservation efforts
- Various modeling methods measure occupancy or abundance data. Abundance data may sometimes be converted to occupancy data however the converse is not typically true.
- Abundance data are frequently amenable to far more sophisticated modeling than are occupancy data. However occupancy data may be superior in some applications, particularly in cases where populations are small or densities low.
- Pollard walks provide a measure of relative abundance rather than estimating absolute population size. The method is employed by numerous citizen science programs and can generate large quantities of data.
- Mark Release Recapture (MRR) provides the most detailed demographic information, and offers the opportunity for the most robust modeling. MRR is the most labor-intensive of the monitoring methods discussed here and carries a risk of damage to sensitive butterfly species during the marking process.
- Distance sampling, like MRR, produces an estimate of population size rather than relative abundance. It does not involve the same risk of damage to butterflies as MRR. Limitations include the requirement for random transects within areas of uniform butterfly density, as well as key assumptions that may not be met in field studies.
- Count circles and field trips can be used as part of citizen science programs to generate large amount of data which, although collected with less standardization than other methods, provide useful information for trend analysis.
- Sightings databases provide another source of data, but dynamics are difficult to track with these data sources.
- There has been a sharp rise in the number of butterfly monitoring programs in North America and a new organization, The North American Butterfly Monitoring Network, is developing resources to make sure data can be effectively captured, managed, shared and analyzed.
- To date, these monitoring data have been underutilized both for scientific research and conservation decision making, but with the advancement of new programs, systems to support them and more widespread knowledge of the program's existence, data use should increase substantially.

Acknowledgements We thank the program directors and volunteers for their dedication in participating in these programs and providing invaluable data to meet research and conservation needs. We particularly thank program managers who shared their meta-data with us so that we could track program activity and compile records across programs. Funding was provided to LR through NSF award DBI-1147049 and from DBI-1052875 through the Socio-environmental Synthesis Center.

References

Bried JT, Pellet J (2012) Optimal design of butterfly occupancy surveys and testing if occupancy converts to abundance for sparse populations. J Insect Conserv 16:489–499

Brown JA, Boyce MS (1998) Line transect sampling of Karner blue butterflies (Lycaeides melissa samuelis). Environ Ecol Stat 5:81–91

Calabrese JM (2012) How emergence and death assumptions affect count-based estimates of butterfly abundance and lifespan. Popul Ecol 54(3):431–442

Collier N, Mackay DA, Benkendorff K (2008) Is relative abundance a good indicator of population size? Evidence from fragmented populations of a specialist butterfly (Lepidoptera: Lycaenidae). Popul Ecol 50:17–23

Dennis RLH, Shreeve TG, Isaac NJB, Roy DB, Hardy PB, Fox R, Asher J (2006) The effects of visual apparency on bias in butterfly recording and monitoring. Biol Conserv 128(4):486–492

Dennis EB, Freeman SN, Brereton T, Roy DB (2013) Indexing butterfly abundance whilst accounting for missing counts and variability in seasonal pattern. Methods Ecol Evol 4(7):637–645

Forister ML, Jahner JP, Casner KL, Wilson JS, Shapiro AM (2011) The race is not to the swift: long-term data reveal pervasive declines in California's low-elevation butterfly fauna. Ecology 92(12):2222–2235

Gross K, Kalendra EJ, Hudgens BR, Haddad NM (2007) Robustness and uncertainty in estimates of butterfly abundance from transect counts. Popul Ecol 49:191–200

Haddad NM, Hudgens B, Damiani C, Gross K, Kuefler D, Pollock K (2008) Determining optimal population monitoring for rare butterflies. Conserv Biol 22(4):929–940

Hanski I, Alho J, Moilanen A (2000) Estimating the parameters of survival and migration of individuals in metapopulations. Ecology 81(1):239–251

Harker RJ, Shreeve TG (2008) How accurate are single site transect data for monitoring butterfly trends? Spatial and temporal issues identified in monitoring Lasiommata megera. J Insect Conserv 12:125–133

Holling CS (ed) (1978) Adaptive environmental assessment and management. Wiley, New York

Isaac NJ, Cruickshanks KL, Weddle AM, Rowcliffe JM, Brereton TM, Dennis RLH, Shuker DM, Thomas CD (2011) Distance sampling and the challenge of monitoring butterfly populations. Methods Ecol Evol 2(6):585–594

Isaac NJ, van Strien AJ, August TA, de Zeeuw MP, Roy DB (2014) Extracting robust trends in species' distributions from unstructured opportunistic data: a comparison of methods. BioRxiv 006999

Kadlec T, Tropek R, Konvicka M (2012) Timed surveys and transect walks as comparable methods for monitoring butterflies in small plots. J Insect Conserv 16:275–280

Koenig WD (2006) Spatial synchrony of monarch butterflies. Am Midl Nat 155(1):39–49

Kuefler D, Haddad NM, Hall S, Hudgens B, Bartel B, Hoffman E (2008) Distribution, population structure and habitat use of the endangered Saint Francis Satyr butterfly, Neonympha mitchellii francisci. Am Midl Nat 159(2):298–320

Link WA, Sauer JR, Niven DK (2006) A hierarchical model for regional analysis of population change using Christmas Bird Count data, with application to the American Black Duck. Condor 108(1):13–24

Longcore T, Mattoni R, Zonneveld C, Bruggeman J (2003) Insect Count Analyzer: a tool to assess responses of butterflies to habitat restoration. Ecol Restor 21:60–61. Available at: http://www.urbanwildlands.org/INCA/. Accessed 15 Jan 2015

Longcore T, Lam CS, Kobernus P, Polk E, Wilson JP (2010) Extracting useful data from imperfect monitoring schemes: endangered butterflies at San Bruno Mountain, San Mateo County, California (1982–2000) and implications for habitat management. J Insect Conserv 14(4):335–346

MacKenzie DI, Nichols JD, Sutton N, Kawanishi K, Bailey LL (2005) Improving inferences in population studies of rare species that are detected imperfectly. Ecology 86(5):1101–1113

Matechou E, Dennis EB, Freeman SN, Brereton T (2014) Monitoring abundance and phenology in (multivoltine) butterfly species: a novel mixture model. J Appl Ecol 51(3):766–775

Matteson KC, Taron DJ, Minor ES (2012) Assessing citizen contributions to butterfly monitoring in two large cities. Conserv Biol 26:557–564

Morton AC (1982) The effects of marking and capture on recapture frequencies of butterflies. Oecologia 53(1):105–110

Murphy DD (1987) Are we studying our endangered butterflies to death? J Res Lepidoptera 26(1):236–239

New TR (1991) Butterfly conservation. Oxford University Press, Oxford, pp 92–100

Nowicki P, Settele J, Henry PY, Woyciechowski M (2008) Butterfly monitoring methods: the ideal and the real world. Isr J Ecol Evol 54(1):69–88

O'Brien JM, Thorne JH, Rosenzweig ML, Shapiro AM (2011) Once-yearly sampling for the detection of trends in biodiversity: the case of Willow Slough, California. Biol Conserv 144(7):2012–2019

Oberhauser KS, Ries L, Altizer S, Batalden R, Kudell-Ekstrum J, Garland M, Howard E, Jepsen S, Lovett J, Monroe M, Morris G, Rendón-Salinas E, Rubino RG, Ryan A, Taylor OR, Treviño R, Villablanca F, Walton D (2015) Contributions to monarch biology and conservation through citizen science: 70 years and counting. In: Oberhauser KS, Nail K, Altizer S (eds) Monarchs in a changing world: biology and conservation of an iconic butterfly. Cornell University Press, Ithaca, pp 13–30

Pellet J, Bried JT, Parietti D, Gander A, Heer PO, Cherix D, Arlettaz R (2012) Monitoring butterfly abundance: beyond Pollard walks. PLoS One 7:41396

Polic D, Fiedler K, Nell C, Grill A (2014). Mobility of ringlet butterflies in high-elevation alpine grassland: effects of habitat barriers, resources and age. J Insect Conserv 18:1153–1161

Pollard E (1977) A method for assessing changes in the abundance of butterflies. Biol Conserv 12:115–134

Pollard E (1982) Monitoring butterfly abundance in relation to the management of a nature reserve. Biol Conserv 24:317–328

Pollard E (1988) Temperature; rainfall and butterfly numbers. J Appl Ecol 25:819–828

Pollard E (1991) Synchrony of population fluctuations: the dominant influence of widespread factors on local butterfly populations. Oikos 60:7–10

Pollard E, Yates TJ (1993) Monitoring butterflies for ecology and conservation: the British butterfly monitoring scheme. Chapman & Hall, London

Pradel R, Hines JE, Lebreton JD, Nichols JD (1997) Capture-recapture survival models taking account of transients. Biometrics 53:60–72

Ricketts TH (2001) The matrix matters: effective isolation in fragmented landscapes. Am Nat 158(1):87–99

Ries L, Mullen SP (2008) A rare model limits the distribution of its more common mimic: a twist on frequency dependent Batesian mimicry. Evolution 62(7):1798–1803

Ries L, Oberhauser K (2015) A citizen-army for science: quantifying the contributions of citizen scientists to our understanding of monarch butterfly biology. Bioscience 65:419–430

Rothery P, Roy D (2001) Application of generalized additive models to butterfly transect count data. J Appl Stat 28:891–909

Royle JA, Nichols JD, Kery M (2005) Modelling occurrence and abundance of species when detection is imperfect. Oikos 110:353–359

Shuey J, Szymanski J (2010) Modified Pollard transects do not predict estimated daily population size for the secretive butterfly, Neonympha mitchellii mitchellii French. J Lepidopterists Soc 66:221–224

Singer MC, Wedlake P (1981) Capture does affect probability of recapture in a butterfly species. Ecol Entomol 6(2):215–216

Swengel AB (1990) Monitoring butterfly populations using the Fourth of July Butterfly Count. Am Midl Nat 124:395–406

Swengel AB (1995) Population fluctuations of the monarch (Danaus plexippus) in the 4th of July Butterfly Count 1977–1994. Am Midl Nat 134:205–214

Taron DJ (1996) Insects. In: Packard S, Mutel C (eds) The Tallgrass restoration handbook. Island Press, Washington, pp 305–318

Thomas JA (1983) A quick method for estimating butterfly numbers during surveys. Biol Conserv 27:195–211

Thomas L, Buckland ST, Rexstad EA, Laake JL, Strindberg S, Hedley SL, Bishop JLB, Marques TA, Burnham KP (2010) Distance software: design and analysis of distance sampling surveys for estimating population size. J Appl Ecol 47(1):5–14

Vandenbosch R (2003) Fluctuations of Vanessa cardui butterfly abundance with El Niño and Pacific Decadal Oscillation climatic variables. Glob Chang Biol 9:785–790

van Swaay CA, Nowicki P, Settele J, van Strien AJ (2008) Butterfly monitoring in Europe: methods, applications and perspectives. Biodivers Conserv 17(14):3455–3469

WallisDeVries MF (2004) A quantitative conservation approach for the endangered butterfly Maculinea alcon. Conserv Biol 18(2):489–499

Zonneveld C (1991) Estimating death rates from transect counts. Ecol Entomol 16:115–121

Developing a Rare Butterfly Database for Conservation Purposes: An Example in Florida Using Citizen Scientists

Dean Jue

Abstract While there are projects building comprehensive databases on changes in butterfly populations over time and space, there is still a need for building a rare butterfly database that can quickly be of value to land managers and decision-makers for whom long-term butterfly population data and trends are not yet available. The chapter provides a frame work for developing a database that can rapidly become useful for butterfly conservation efforts and decision-makers at different geographic scales and is based upon 6 years of developing a statewide rare butterfly database within Florida using citizen scientists. Topics covered include funding, staffing, long-term database support, recommendations for building an effective rare butterfly database, and organizational considerations in maximizing the effectiveness of project volunteers. The chapter relates each of these specific topics to the real-world experiences in developing Florida's rare butterfly database and discusses the considerations, tradeoffs and rationale for the decisions that were made. This chapter is likely to be of special interest to zoo and museum directors starting up a butterfly conservation project, land managers, and government agencies.

Introduction

Butterflies are iconic insects that have become popular with the general public as well as with researchers. With this heightened interest has come increased demand for land management practices that are butterfly-friendly as well as use of the legal statutes such as the United States (U.S.) Endangered Species Act to protect species that are in serious decline (Lundstrom 2014). Although the decline of a flagship species such as the Monarch (*Danaus plexippus*) has already been well-documented, such extensive research and data are usually nonexistent for the vast majority of butterfly species that may be in decline.

D. Jue (✉)
Florida Natural Areas Inventory, Tallahassee, FL, USA

Florida Resources and Environmental Analysis Center, Florida State University, Tallahassee, FL, USA
e-mail: DJue@admin.fsu.edu

© Springer Science+Business Media B.V. 2015
J.C. Daniels (ed.), *Butterfly Conservation in North America*,
DOI 10.1007/978-94-017-9852-5_4

One of the important pieces of evidence for conservation action for a butterfly species is data that its population is declining and being negatively impacted. In order to provide these data, there must be baseline data available on a species' past distribution and population numbers relative to the current situation. There are several global projects that are building such comprehensive databases across all butterfly species within particular countries. However, there is also a need for a more focused approach to building a database that can quickly be of value to butterflies that are already believed to be in serious decline for which long-term population data and trends may not yet be available.

This chapter provides a frame work that utilizes a focused approach to developing a database that can rapidly become useful for butterfly conservation efforts and decision-makers at different geographic scales, from a local and perhaps personal level to a database that could potentially serve as a national butterfly conservation database. The paper is based upon 6 years of developing a statewide rare butterfly database for Florida from the Florida Fish and Wildlife Conservation Commission's (FL FWC) State Wildlife Grants (SWGs) funding. The overall framework and discussion should be applicable to other situations on a global scale, although the illustrative project being discussed is North America-centric. Some specific audiences for which this chapter should have relevance include

1. zoo and museum staff starting up a butterfly conservation project;
2. land managers wanting to recruit citizen scientists and volunteers to help inventory the butterfly fauna on lands they manage;
3. government agencies responsible for the management of biological resources that want to start building a rare butterfly database for the geographic area under their jurisdiction.

Why Build a Rare Butterfly Database?

Butterfly monitoring using citizen scientists began in the United Kingdom (UK) almost 40 years ago and has expanded to include a number of other European countries (Butterfly Conservation Europe 2014). With such long-term data available, it has been possible to assess the need for butterfly conservation, the relationship of butterfly populations to land management, or to utilize butterflies as indicator species (van Swaay et al. 2008).

Butterfly monitoring projects in other areas of the world are more recent or have been less organized. In Australia, China, and Israel, efforts in those directions began in the late 2000s (Arnhold 2009). Although there are a number of butterfly monitoring projects in North America, the coordination of such projects on a nationwide scale began only in 2012 and the vast majority of such projects started in the 1990s or later (North American Butterfly Monitoring Network 2014). A detailed overview of the various butterfly monitoring programs in North America was presented in the previous chapter.

While all of these North American butterfly monitoring programs and associated databases have valid goals and are worthwhile, none of them is oriented specifically toward collecting and monitoring rare butterfly distribution and abundances in a systematic manner. Many contain abundance data of primarily the more common butterfly species (e.g., North American Butterfly Association 2014). Most do not reflect a systematic search for a rare butterfly species within a particular geographic area. There is also usually a limited capability to record the absence and non-occurrence of a species in an area after an extensive search for it within that area. And, unlike Europe, the timespan of the butterfly abundance and distribution data is still relatively spotty in distribution and usually covers a relatively short timeframe of less than 20 years.

Despite the lack of long-term population trend data for even most of the common North American butterfly species and the incompleteness of data for the rare and declining ones, many government agencies are increasingly being asked to consider rare butterflies and their conservation needs. Currently, for the majority of projects and agencies in North America that have a immediate need for a rare butterfly database to be used for land management decisions or for species recovery, the most expedient option is likely to be for the associated agency and its personnel to build the rare butterfly database themselves.

Resources for Database Development

If a rare butterfly database is needed, it is important to evaluate the available resources for developing it. Resources can be broadly grouped into the categories of funding, staffing, and long-term database support and maintenance. If existing resources are deemed inadequate to support building the rare butterfly database at the level of detail and functionality required, an effort may be needed to supplement the existing resources through additional grant funding or donors.

Funding

Funding is often the resource over which an individual or agency developing a rare butterfly database has the least amount of control. An agency is provided with a fixed amount of money to complete this task; a nonprofit organization takes on the conservation of a rare butterfly species as a high-profile project and needs to find additional funding sources or raise more money to develop the project's database, or a staff member has received a fixed price grant to develop a butterfly database from a philanthropic organization.

If additional sources of funding are necessary for optimal development of the butterfly database, other possible options could be pursued:

1. Apply for additional funding through grants that may be available from other public agencies or philanthropic organizations. Depending on the initial funding

source, the existing available funds may be able to be used as cost-share or match in order to help procure additional funding.
2. Partner with another agency. If a partnering agency can perform some of the required work using its staff or resources, this will free up some of the available funding to conduct other work that may be required in the database development process.
3. Some agencies (e.g., state parks, museums, zoos) may have Friends groups. If so, work with the Friends group to develop fundraising campaigns or utilize their help in performing some required database development tasks.

Staffing

Ideally, the staff necessary for collecting, developing, and maintaining the rare butterfly database was identified during the early stages of the planning process and funds are available to utilize those staff members for the anticipated project life cycle. Should staff funding be inadequate, there are two possible sources to supplement agency staff that may be low-cost as well.

The first source is students at schools and universities, perhaps especially at the graduate school level. Students who are exploring career options may be willing to participate in a real-world project providing them with on-the-job experiences. Possible roles for a student in developing a rare butterfly database range from field surveys to data entry to database design to web application development, depending on the particular needs of the project and the skill set of the student. Student levels could probably range from a mature high school student to a graduate student seeking a project for a master's or doctoral degree.

The second source for supplemental staffing is citizen scientists or retirees who are looking to participate in worthwhile projects. Some of these potential recruits may already have a scientific background. There are already citizen scientists throughout many parts of the U.S. that have contributed substantially to new information about the distribution and status of rare butterflies in the U.S. by surveying for butterflies in areas that have been historically under-surveyed. The case study discussed below relied heavily on the involvement of citizen scientists in developing Florida's rare butterfly database.

The potential problem with these supplemental staffing options is that if the people if they are not actually on the project's payroll there is a limit to what can be done about non-performance of their expected responsibility. They can be dismissed, of course, but that doesn't solve the more immediate problem of their non-performance. One way to help minimize this potential problem is to provide volunteers with a list of their expected responsibilities and duties upon their involvement with the project. This list, of course, can vary from one individual to another but it will help reduce the likelihood of misunderstandings creating problems for the project in the future. Another way is to provide some remuneration to these individuals for their involvement (e.g., access and training on sophisticated equipment, reimbursement for gas mileage for field work).

Long-Term Database Support

A critical question to be addressed is where should the rare butterfly database reside and ultimately be maintained. This is not as simple as saying that it will reside within the agency that developed the database, although that is probably a likely default answer. Specific alternatives that should be considered before accepting the default answer are listed below:

1. If 24/7 access to the database is required by a number of participating partners, might the database best reside in the cloud? If the database is to reside in the cloud, have all the security issues been resolved relative to possible disclosure of rare and endangered species locations if the cloud environment is hacked?
2. If the database is going to be used in objective and non-partisan decision-making, might maintaining the database in the developing agency itself compromise the apparent objectivity of that database because it is viewed as serving the agency's mission rather than simply being an objective and fair research or conservation database?
3. If the database is to be a statewide rare butterfly database might the database best reside and be maintained with other databases in the state's natural heritage database?

Some of the above issues could be addressed through serving up the rare butterfly data to external data-requesting agencies using web services by the database-hosting agency.

Building an Effective Butterfly Database

The building of a rare butterfly database for conservation purposes takes time, money, and staff. But the ultimate hope, of course, is that the database will be effective and make a difference in helping to conserve rare butterfly species. There are several considerations in the design and building of the database that will help ensure it will make such a difference to both the developing agency itself as well as outside users. These points are discussed here.

Defensible Assessment and Ranking of Butterfly Status

When building a rare species database, one of the attribute fields is usually a ranking or status field that categorizes a species' degree of rareness. It may be a tri-level status (e.g., endangered, threatened, or species of concern), such as used by the U.S. Fish and Wildlife Service (2014), or it may assign status ranks to different levels of geography (e.g., global, state) as developed by NatureServe (2014b). Some states' fish and wildlife agencies have their own state-wide ranking system for rare species.

When developing the rare butterfly species database, you should either adopt one of the existing commonly used species ranking systems or develop your own if you believe that none of the existing ones suits your purposes. If you develop your own, keep in mind that you will need to explicitly define many of your terms (e.g., "rare") and may also need to state how it differs from other ranking systems that utilize that same term.

For maximum utility and ease of comparability of your database, the adoption of a ranking system that has been vetted and is being used by other agencies or organizations probably makes the most sense. For example, a statewide rare butterfly databases should probably adopt either the ranking system used by its state's fish and wildlife agency or that used by their state natural heritage program.

Unless a single-species butterfly database is being built, one of the high priority tasks that should be completed after the adoption of the ranking system is an initial assessment and ranking of all the rare butterfly species to be included in the database. This ranking forces a systematic review of many if not all of the butterfly species in the geographic area of interest. This will help ensure the completeness of the database to be built as well as provide an initial baseline status and ranking. These statuses and rankings will change over time as data are collected. In the meantime they will help identify and prioritize species which are believed to be in the greatest need for immediate help.

How is this initial assessment and ranking to be done? There are at least three different ways, with the best way dependent on the specific situation for the organization that is developing the database:

1. Have the assessment and status reviews done by an in-house expert or consultant;
2. Send species list out for review to a group of experts as well as knowledgeable citizen scientists and then assign a composite ranking; or
3. Utilize a pre-existing ranking and assessment for each species. For example, NatureServe (2014c) has global rankings for the butterfly species in the U.S.

Verifiable Records

In order for the database to be widely accepted outside of the developing agency or organization, the database records must be verifiable. Because the database is focusing on rare species, its records need adequate documentation to prove, as a minimum, that the species was correctly identified. Although it is tempting to accept all sight records from today's recognized authorities, there is a good probability that in 100 years very few people will recognize the name of those authorities.

Verifiable records mean digital photographic documentation as a minimum for rare butterfly species that can be diagnostically identified from that photograph. Voucher specimens may be required in some instances for rare taxa that are difficult if not impossible to distinguish from other taxa based on just digital photographs. This topic is discussed in more detail in the "Organizational Considerations" section of this chapter.

Use in High-Level Decision-Making

The ultimate proof of whether the rare butterfly database is effective will be if it makes a difference in the actions, plans, or policies of higher-level decision makers. If the database has been built for helping to recover the population of a single species, that higher level decision maker could be the same person that developed the database. Or that higher level may be a committee, a state conservation agency, a national fish and wildlife agency, or even an international consortium of countries.

Although not an absolute prerequisite to being an effective database, a characteristic that is desirable for most databases is the ability to be easily integrated and combined with data from other databases. This minimizes the duplication of efforts and allows a rare butterfly database to focus its data collection on rare butterfly data without the need for collecting related data of interest (e.g., land use cover, rainfall pattern, remote sensing data, existing and proposed conservation lands boundaries).

Organizational Considerations for Building the Database

When building the database, there are a number of different paths that can be taken to populate it. The optimal path will depend on a number of factors. These include the anticipated size of the database, the number of species that will be included in the database, the amount of funding available, the number of internal staff available for helping, and the hardware and software environment that will be used for implementing the database.

Although there is no one way to build the database, there are some commonalities that are likely to be encountered while developing the rare butterfly database. A vetting of the various options prior to actual project commencement will help in efficiently developing a consistent and useful butterfly database.

Data Coordinators

If the geographic scope that the rare butterfly database will cover is large, it may make sense to sub-divide the geographic area covered by the database into smaller areas and have a regional data coordinator for each of the regions. For databases and projects covering small geographic regions focusing on just a very few species, regional data coordinators may not be necessary.

The rationale for having regional data coordinators is that the coordinators know their own local areas, butterfly fauna, and other potential local butterfly surveyors better than the database developer or principal investigator who may be hundreds of kilometers away. A list of tasks that could fall naturally to a regional data coordinator includes the following:

1. Recruit other project participants;
2. Serve as contact person for project participants from the region;
3. Identify lands within the region to survey for the rare butterflies of interest;
4. Coordinate the surveys within the region by project participants to ensure optimal coverage and collection of data with minimal overlapping of data and surveys; and
5. Review and correct as needed the regional data prior to submission to the project's database administrator

Of course, there above list is just an example of the possible tasks and will need to be customized to the particular needs and situations of each butterfly database.

Having regional data coordinators also provides a better ability to monitor butterfly phenology patterns and emergences. Although there are already data for the number of broods and the associated flight time for each brood in the scientific and popular literature, local and regional weather variations can affect the timing of eclosures and peak abundances by several days to several weeks. Each regional coordinator can closely monitor the local and regional weather variations and time the data collection surveys to coincide with periods when the abundance for a targeted rare species is likely to be the highest.

Planning Meetings

For databases where regional data coordinators make sense, it is also important to make sure that all the regional data coordinators are kept abreast of database needs and project priorities. For example, if a targeted rare species is found in greater than anticipated abundance in six out of eight regions, the next year's project objective may be to target an alternative species among the eight different regions.

This type of data coordination among the different regional data coordinators can best be done in a planning meeting involving the data coordinators and the project investigator and manager. The frequency of such meetings can be left up to the discretion of the project manager, but one or two meetings per year is probably a reasonable number.

At these meetings, topics of discussion and activities could include the following:

1. Overall problems and possible solutions relative to the project as a whole;
2. Unique problems that may be faced by one or two regions and possible solutions;
3. Review of the field work conducted and data collected by each region since the last meeting;
4. Goals and objectives relative to the project for the next year; and
5. Discussion of specific species problems, if any, and review of voucher specimens for potential problem species.

Again, there could be additions or deletions to this list of possible tasks.

Standardization of Data

All data entered into the rare butterfly database should be standardized if possible. Each record should have required attribute data fields filled with relevant data, and those data should be comparable across the different records within the database. The potential for standardized data today is much better than 20 years ago. Stand-alone applets with pull-down menus can be written for mobile field devices that will help ensure consistency across data records.

Even with today's technology, however, there will probably be rare butterfly records that should be part of the database that are missing some desired data (e.g., historic data). Ways to minimize such problems include the following:

1. Identify a minimal set of data attribute fields required for entry into the database. A likely minimal set would include who saw what butterfly at what time at which location. If any one of these four data elements is missing, the record will not be entered into the database;
2. Have a required attribute field that indicates the accuracy of the data record. For example, a locational accuracy field can indicate whether the location is based on a global positioning system (GPS) unit, vehicle mileage from a known landmark, estimated from a Google Map on the computer, or from a United States Geological Survey quadrangle sheet;
3. Develop a small additional set of desired data fields that are deemed most vital for the database. Examples of such attribute fields could be presence or absence of the host plant in the vicinity, estimated long-term viability of site, or potential problems at site; and
4. Although most data would ideally be captured on-site on an electronic device and transmitted in real-time to the remote butterfly database, there will be some percentage of the data that will be recorded and entered manually (i.e., from paper). Developing a simple one-page form that can be used to record both the minimal and the additional desired attribute data set can help facilitate the database project in efficiently entering the desired information for all records.

Partitioning of Data Collection Surveys

For databases covering larger geographic areas, it makes sense to have regional data coordinators. By the same logic, the data collection surveys could be partitioned along geographic lines similar to those defined by and assigned to each of the data coordinators. The advantages are listed below:

1. This helps force a more comprehensive coverage of each region for rare butterfly surveys. There is a tendency to return to known localities for certain butterfly species because the butterflies are easiest to find there. But the flip side is that there may be a lack of surveys for a particular rare butterfly species in other areas.

Dividing larger areas into smaller regions will help make project participants look for rare butterfly species in their own region rather than in the traditional spots that may be many kilometers away;
2. This will lead to a more comprehensive survey of potentially important butterfly areas within each of the regions;
3. It may be easier to recruit citizen scientists to help with the database development effort because some volunteers may prefer to survey in just a very local area that they may already visit and have no desire to survey outside of that one area.

Collecting vs. No Collecting

One issue pertaining to rare butterfly database development that may arise is that of collecting versus non-collecting of voucher specimens, especially if outside helpers such as citizen scientists are involved. Some people are opposed to collecting voucher specimens and may not wish to participate in projects that involve collecting, especially if the butterflies are rare.

Answering the following questions may clarify the need for voucher specimens when developing a rare butterfly database:

1. Are the butterfly taxa for the database identifiable to the correct taxon from just digital photographs or is a voucher specimen necessary for positive identification (e.g., for genitalia examination)?
2. Is there any need for additional taxonomic work on the species?
3. Does your institution or agency have the proper facility and/or relationship with an off-site museum for long-term cataloging and storage of voucher specimens?
4. Do you have the permit required to collect voucher specimens at the site? Many federal agencies and some state agencies require a collecting permit for such purposes.

The arguments against collecting voucher specimens usually center on a general opposition to killing living organisms and the belief that the taking of rare butterflies is pushing their population levels closer to extinction.

Ultimately, the database developer must weigh the potential benefits of possible wider citizen scientist participation with a no-voucher specimen collecting policy versus the necessity of voucher specimens to accomplish the goals of the rare butterfly database. Some explicit project guidelines and policies that may increase participation in the project by individuals who are generally opposed to collecting include the following:

1. Voucher specimens will be collected only if absolutely necessary for positive taxon identification or anticipated taxonomic studies and will follow the guidelines of the Statement on Collecting by The Lepidopterists' Society (2014);
2. Diagnostic digital photograph is the preferred method for documenting all butterfly species whenever possible.

3. Individuals opposed to collecting of voucher specimens will not be expected to collect any specimens. If warranted, however, another project staff member may be sent out to attempt to collect voucher(s).
4. Location data associated with each database record will be provided only to those whose requests adequately document their need to know (e.g., planners, conservation organizations). Location data will not be released to private collectors.

Creating and Maintaining the Database

It is often easy to focus on the end product, a database containing rare butterfly locations and associated data that are useful for conservation planning purposes, and lose sight of the steps required to get to that end point. There are critical but mundane tasks such as data entry and database maintenance that are required to realize a rare butterfly database.

In today's web-based computing environment, it is often assumed that primary data entry can be done by field workers using mobile devices that upload their digital data directly into the database. While this is an increasingly common scenario, it downplays the need for quality checks of the primary data prior to its incorporation into the database. It ignores the potential complications of incorporating relevant historic data into the database. There also will undoubtedly be data provided to the database project which were not collected using mobile field devices but provided in paper format or an e-mail. None of the issues associated with these alternative data entry methods are insurmountable but they do require extra resources and personnel to address them and such additional costs would ideally be incorporated into the original project cost estimate.

To avoid the likelihood of such unexpected costs placing a financial burden on the project, the following questions should be considered prior to finalizing the estimated budget costs for developing and maintaining the rare butterfly database:

1. How much historic data will be entered into the database? Is any of it already in digital format or will it require on-site visits and manual transcription of field notes and specimen tags?
2. How will newly-collected data be entered into the database? Do you have sufficient staffing to enter the paper field notes and standardized data forms into the database? If you are collecting data using mobile field devices, have you accounted for development, troubleshooting, and any required customization of the applets across the many mobile platforms that may be used (e.g., Android, iPhone, tablets, GPS devices)? Who is performing the quality control checking of the new data and how will it be done?
3. What is the estimated cost and personnel requirement for maintaining the rare butterfly database after the initial funding for the database development ends? There will undoubtedly be updates to existing records (e.g., recent surveys

suggest the butterfly no longer occurs at this location) as well as submission of new locality records, all of which will require quality control checking as well. Requests for updates to existing applets may also occur.

Integration of Rare Butterfly Database with Other Data Sets

The purpose of the rare butterfly database is for it to be used. Its primary user may be the agency or organization responsible for its initial development, but another measure of the database's success is the value it provides to outside organizations. Also, the butterfly database may need to access data sets that have been developed and being maintained by outside organizations.

Consequently, it is important to identify other data sets and databases that will be integrated with the butterfly database and make sure that the ability to exchange and use data among all these identified data sets is as transparent as possible. An example of this might be the consistent and standardized usage of the same format for the date field.

An important related data set for a regional or state butterfly conservation database is a conservation lands database. That database should contain the boundaries of all conservation lands at all governmental levels within the region covered by the butterfly database as well as any conservation lands that may be privately-owned. This relevant data set will show the degree to which known localities for a rare butterfly may already reside on conservation lands and thus be reasonably well-protected as compared with butterfly species for which there are no known localities on conservation lands. This type of information can help prioritize butterfly species conservation efforts.

Case Study: Florida

The above discussion on developing a rare butterfly database is based on the experience of doing so in Florida in the United States. This section describes the hows and whys of decisions made during development of the Florida rare butterfly database in hopes that this information will help other agencies and organizations better visualize the options and tradeoffs that may need to be considered while developing their own rare butterfly database.

Background

The Florida Natural Areas Inventory (FNAI) is one of 82 natural heritage programs operating in North and Latin America under the NatureServe umbrella (NatureServe 2014a). The FNAI maintains a statewide database on the locations and conditions of rare and at-risk plants, animals, and ecological communities. In addition, FNAI maintains comprehensive data on the boundaries, ownership, and management status of all lands within the state managed for conservation.

Until the mid-2000s, FNAI's database and research efforts were skewed towards plants, vertebrates, and natural communities. At that time, the FNAI rare species tracking list included only nine taxa of butterflies. The FNAI began to remedy this situation in 2005 when it received a Florida Fish and Wildlife Conservation Commission (FL FWC) State Wildlife Grant (SWG) to focus specifically on improving the breadth and completeness of its invertebrate database. FNAI staff requested feedback from Florida's professional lepidopterists and preeminent butterfly watchers to determine which butterflies should be added to the FNAI tracking list and what state ranks should be assigned under the NatureServe ranking system, with S1 being the rarest and most threatened, and S5 being the most common and secure species. As a result of that effort, the FNAI augmented the number of tracked butterfly taxa to 78 taxa in Florida. This augmentation of FNAI's tracking list for butterflies provided the first step towards building an effective database – a defensible assessment and ranking of the current butterfly statuses specific to Florida.

Strategy for Development of FNAI's Rare Butterfly Database

With completion of the expanded butterfly list, the next step was to accurately populate the FNAI database with current locations and statuses for as many of the newly-tracked butterflies as possible. The FNAI applied for another FL FWC SWG that would focus specifically on populating its statewide database with current locations for the newly-tracked butterfly taxa on conservation lands. The FL FWC provided 6 years of funding for this effort through SWGs that began on 1 July 2007.

The funding allowed the FNAI to focus staff and resources specifically towards building the rare butterfly database. Since FNAI is the natural heritage program for Florida and its data are already considered in the state's environmental land acquisition and restoration process, it made sense for the new database to reside within FNAI, especially since FNAI had the necessary staff and computing resources to host the database.

FNAI researchers planned for a very active role for citizen scientists with the SWG. This stemmed from necessity, since the grants required a 50 % cost-share. All FNAI employees are soft-funded by specific projects, so their work time could not be used to meet the SWG cost-share requirement. But using knowledgeable citizen scientists who could survey for the FNAI-tracked butterfly species could provide the required cost-share.

In addition to the necessity of involving volunteers to satisfy the grants' financial requirements, there were two practical reasons for having citizen scientists participate in the project. First, Florida is a large state that stretches over almost 1,300 km (800 miles) from Pensacola in the northwest to Key West in the extreme south. It would be difficult to adequately survey for the many butterfly taxa of interest using only FNAI staff. Projects that span large spatial extents, such as FNAI's grants to inventory rare butterflies statewide, are especially suitable for the involvement of citizen scientists (Devictor et al. 2010).

Secondly, there was already a sizeable network of people in Florida who were regularly conducting butterfly surveys on their own and already knew many locations to look for rare butterflies within their particular area of the state. Rather than duplicate their knowledge and efforts, it made more sense to coordinate and organize their frequent butterfly surveys in a way that the information they were collecting could be entered into the FNAI database. The key to making such a citizen science effort successful was to develop an organizational structure that minimized duplication of effort among the participants while ensuring the accuracy and quality of the collected data.

The FNAI utilized regional coordinators to organize and coordinate the data collection and field surveys within different regions of Florida. The regional coordinators were volunteers who donated their time and expertise and in return got reimbursed for their car mileage while conducting the field surveys. All of the regional coordinators had been involved with butterfly surveys in Florida for 5 or more years. Consequently, there was minimal need to provide formal training in the identification of the rare butterfly species to the group.

To coordinate the survey efforts statewide, the regional coordinators met annually in Gainesville, Florida, to coordinate strategic field surveys across the different regions and to share experiences, observations and suggestions. They were joined by other interested butterfly experts and scientists who provided support for the project but did not have time to take a more active role. Surveys were either conducted by the regional coordinators themselves or by other knowledgeable volunteers recruited by the coordinators.

Field surveys focused on finding new populations of the rarest Florida butterfly species on conservation lands. A standardized data form was used to collect information about each survey. Regional coordinators were provided with a GPS unit to obtain geographic coordinates for sightings. A diagnostic digital photograph was required by FNAI for documenting the occurrence of a tracked butterfly species at a new location. Sightings and reports were considered "questionable" unless this confirming photograph was obtained. Photographs of species that were difficult to identify were sent to butterfly authorities in Florida for confirmation.

Survey data for FNAI-tracked butterfly species were then entered into the FNAI database using the Biotics 4 data management software. Biotics 4, which was developed by NatureServe for the natural heritage programs, utilizes an Oracle database for managing tabular data and a customized ESRI ArcView geographic information system software for managing and mapping the spatial component of the data. FNAI's statewide Biotics database includes occurrences of not only butterflies but all FNAI-tracked plants and animals as well as exemplary natural biological communities and other noteworthy natural features (e.g., special bird breeding areas). All these data, as well as FNAI's conservation lands map layer, can be used during statewide decision-making processes for environmental land acquisition and management.

Relative to the earlier discussions in this paper, the FNAI implementation strategy for its rare butterfly database incorporated

1. verifiable records through requirement of digital photograph to document new locations for the rare butterflies;
2. data coordinators;
3. planning meetings;
4. standardization of data through its field data form;
5. partitioning of surveys among each of the regional coordinators; and
6. integration of the rare butterfly database with other data sets.

The issue of collecting versus no collecting was easy to resolve in this particular instance because all of the volunteers were members of the North American Butterfly Association (NABA), which has a policy of non-consumptive enjoyment of butterflies (i.e., no collecting). The no-collecting policy was easy to implement in Florida because there are almost no butterfly species in Florida that cannot be diagnostically-identified with a good photograph.

Making a Difference

The creation of the rare butterfly database has made a difference both in the knowledge about the current distribution and status of the rarer FNAI-tracked butterflies but also in management actions toward helping to maintain and conserve populations of those rare butterflies. Specific examples include the following:

1. The number of element occurrences (i.e., distinct populations) of rare butterflies in the FNAI database increased from 33 in 2007 at the start of the rare butterfly database development to 396 in 2013;
2. The discovery of a new butterfly taxon in Florida, the Brown Elfin (Calhoun et al. 2009);
3. The use of the survey data collected from the FL FWC State Wildlife Grants to develop land management guideline plans for five of the rarest Florida butterflies;
4. The use of the survey data in identifying critical populations of Florida butterflies and working with land managers to adjust their land management practices to help ensure the long-term viability of those populations.

Summary and Key Points

This chapter discusses many of the considerations pertaining to development of a rare butterfly database by land management agencies, agency staff, conservation organizations, or other non-profit organizations. Citizen scientists and other volunteers can play a major role in helping to populate such a database while reducing the direct costs to the developing agency. An important consideration will be to ensure that data collection and documentation are standardized to enable comparability and usability of the data. Another important aspect for a successful implementation will be accommodating the wishes and desires of those volunteers and citizen scientists while minimizing compromises to the agency's data standards.

The implementation of a statewide rare butterfly database by the Florida Natural Areas Inventory is provided as a case study that successfully addressed the requirements of working with citizen scientists in developing its butterfly database. There is no one implementation model that fits all, however, and agency flexibility will probably remain a key to other successful endeavors.

Acknowledgements Funding was provided to FNAI for the statewide butterfly surveys by the Florida Fish and Wildlife Conservation Commission. The volunteers who served as regional coordinators for the grants were B. Berthet, B. and L. Cooper, B. DeWitt, A. Edwards, M.A. Friedman, S. Jue, S. Koi, K. Malone, E. Nuehring, and D. Stillwaugh. Professional lepidopterists who met with the coordinators group at the annual meetings in Gainesville were J. Calhoun, J. Daniels, and M. Minno. Without the dedication and involvement of all these participants the augmentation of the FNAI database with so many butterfly records would not have been possible.

References

Arnhold T (2009) Now butterflies are also being counted in China, Australia, and Israel. Public Release, 14 Aug 2009. http://www.eurekalert.org/pub_releases/2009-08/haog-nba081409.php. Accessed 15 Sep 2014

Butterfly Conservation Europe (2014) Butterfly monitoring. http://www.bceurope.eu/index.php?id=339. Accessed 17 Sep 2014

Calhoun JV, Friedman MA, Slotten JR (2009) Elfin magic: a new Florida State butterfly record. S Lepidopterists News 31:1–2

Devictor V, Whittaker RJ, Beltrame C (2010) Beyond scarcity: citizen science programmes as useful tools for conservation biogeography. Divers Distrib 16:354–362

Lepidopterists' Society (2014) Statement on collecting. http://www.lepsoc.org/statement_on_collecting.php. Accessed 10 Sep 2014

Lundstrom J (2014) Petition seeks to protect monarchs. http://www.xerces.org/2014/09/05/20431/. Accessed 22 Sep 2014

NatureServe (2014a) A network connecting science with conservation. http://www.natureserve.org. Accessed 14 May 2014

NatureServe (2014b) Conservation status assessment: identifying threatened species and ecosystems. http://www.natureserve.org/conservation-tools/conservation-status-assessment. Accessed 4 Sep 2014

NatureServe (2014c) NatureServe Explorer. An online encyclopedia of life [web application]. Version 7.1. NatureServe, Arlington. http://explorer.natureserve.org. Accessed 4 Sep 2014

North American Butterfly Association (2014) Butterflies I've seen. http://www.nababis.org/home/Index. Accessed 2 Sep 2014

North American Butterfly Monitoring Network (2014) Welcome to NAB-NET. http://www.nab-net.org. Accessed 2 Sep 2014

U.S. Fish and Wildlife Service (2014) Endangered species glossary. http://www.fws.gov/endangered/about/glossary.html. Accessed 4 Sep 2014

van Swaay CAM, Nowicki P, Settele J, van Strien AJ (2008) Butterfly monitoring in Europe: methods, applications, and perspectives. Biodivers Conserv 17:3455–3469

Butterfly Conservation Genetics

Emily Saarinen

Abstract Conservation programs are challenged with preventing biodiversity loss; losses that have typically been managed as preservation of individual species and groups. An increasing focus on different levels of biodiversity has promoted methods evaluating the genetic diversity in these species as a tool in preventing species loss. Therefore, most conservation practitioners today realize the need to evaluate, preserve, and manage population-level and range-wide genetic diversity as an essential component in the preservation of species. Recent methodological (e.g. non-lethal tissue sampling) and technological advances (e.g. high-throughput sequencing) have greatly increased our ability to evaluate genetic diversity in endangered and threatened butterflies. In this chapter, we provide a framework and guide for conducting genetic research and for including it in the management plans of endangered Lepidoptera. We discuss the variety of molecular markers and tools available to researchers as well as guidance in implementing them. We provide a step-by-step guide for conducting tissue sampling, evaluating statistical concerns, and provide advice and troubleshooting tips. We review several case studies that have successfully used molecular data and incorporated the results into management plans. The chapter concludes with a summary of techniques, future directions, and lessons learned from conservation genetics study of Lepidoptera of conservation concern.

Introduction

This chapter is meant to be used as a guide for practitioners interested in including genetic data in their conservation and management plans. Here, I use the term "practitioner" to include academic scientists working with wildlife managers, state and federal agents working with or without academic support, and anyone who may need to interpret genetic data as it pertains to butterfly conservation. I especially wish to clarify misconceptions related to molecular work and provide useful advice for practitioners as they consider the utility of genetic data. I often see

E. Saarinen, Ph.D. (✉)
Division of Natural Sciences, New College of Florida, Sarasota, FL, USA
e-mail: esaarinen@ncf.edu

© Springer Science+Business Media B.V. 2015

J.C. Daniels (ed.), *Butterfly Conservation in North America*,
DOI 10.1007/978-94-017-9852-5_5

recommendations to "include genetic studies" or "perform molecular analyses" in management plans or proposals. These studies, if done poorly, are very expensive, time-consuming, and may not provide the answers to the genetic questions required for management planning. They may also be called into question from a legal perspective and, if done poorly, may even delay conservation action (Morin et al. 2010).

Another concern in genetic studies is that they are performed simply because we have the technology to do so (we do it because we can!). Again, gathering data of any kind is expensive and molecular data is usually considerably expensive for non-model organisms (including those that are threatened or endangered). In this chapter I provide a series of steps that can be used to help guide practitioners in making wise choices regarding the use of molecular data and the types of questions that molecular data can be used to address.

In this chapter, I provide background material about types of molecular markers and statistical concerns. I then provide a step-by-step plan for implementing a conservation genetics program. I highlight case studies where genetics have been successfully incorporated into management plans with useful outcomes. These case studies were chosen to highlight the decision-making process and show effective results. I end the chapter with a summary of lessons learned from butterfly conservation genetics initiatives and discuss future directions and uses of molecular methods on the conservation of butterfly taxa.

What Is Conservation Genetics?

Conservation genetics as a discipline is not very new, but its importance in management planning has become increasingly useful in recent years (Amos and Balmford 2001). As early as the 1970s, Sir Otto Frankel began writing of the importance of genetics in conservation of species. The father of conservation genetics is typically credited to Michael Soulé, and his ideas on the importance of maintaining genetic diversity to conserve species has been expounded by countless other researchers, notably Avise (1994), Frankham (1995), and DeSalle and Amato (2004). These pioneers helped highlight the mechanisms of why genetics are important in conservation. The overarching theme is that maintaining genetic diversity will help maintain species-level diversity (and that species-level diversity is imperative for higher-level ecosystem functioning and maintenance of ecosystem services) (Costanza et al. 1997). Put simply, genetic diversity provides organisms with the fuel to adapt to environmental change.

Conservation genetics questions may be focused in many of several scales; at the species, population, and individual levels. Conservation genetics questions include but are not limited to:

- Addressing taxonomic utility and determination of management units
- Hybridization and genetic introgression between populations/species

- Detection of novel genetic variation/variability
- Detection of levels and losses of genetic variability
- Connectivity between populations
- Inbreeding
- Loss of fitness due to loss of genetic diversity
- Captive breeding and propagation programs
- Invasion of a population by a pathogen
- Decline in population size as detected by genetic methods
- Illegal wildlife trade of whole and partial individuals, and, most recently
- Questions of assisted migration/translocation in the face of global climate change

Molecular Markers: What Are They and How Do I Pick the Best One for My Project?

There is a great jargon-filled literature base when it comes to molecular markers. Acronyms and abbreviations are everywhere, and some of these are redundant! It is no wonder that practitioners often pale at the discussion of molecular methods at the outset. This section is meant to decode the jargon and to provide a useful way to proceed with selecting the best markers to answer your genetic questions given your agency/institution budget and position.

Many academic scientists have been working in the field of ecological genetics with Lepidoptera, however, these projects are not typically focused on questions of direct conservation concern or evaluations of threatened taxa (Vernesi et al. 2008). Interesting molecular approaches have included quantitative genetic evaluations of adult characteristics (e.g. eyespots) (Brakefield 1998) and research using enzymes and enzymatic variation as a diagnostic tool for species delineation (Goldstein et al. 2000) and recognition of hybrid areas (Scriber et al. 2008). Other enzyme studies have tried to find protein forms that may be under selection and therefore have a fitness component. Research advances in this area (proteins under selection) do have relevance in the conservation world as evidenced by the long-term study of *Melitaea cinxia* in Finland and the near-shore islands. Evaluation of the *Pgi* (*phosphoglucose isomerase*) locus has shown that different allelic forms impart varying levels of dispersal ability in adult butterflies (Wheat et al. 2010). In this section, I describe several of the most commonly used molecular markers in butterfly research.

Allozymes

Allozymes are one of the earliest forms of genetic data in conservation programs. Allozymes are alternate forms of a protein (coded by a single locus/gene) and therefore represent data related to traits that provide different levels of fitness in an organism. Allozymes can be extracted from an individual (typically a whole, freshly-killed specimen), processed, and visualized on an agarose gel. Many undergraduate and graduate research students find a point of entry into conservation biology by evaluating allozymes because they are easy and relatively inexpensive to evaluate. Working with allozyme data does have several downsides. First, the proper isolation of allozymes must be done from relatively fresh tissue, typically from the thorax or abdomen of the butterfly. This is therefore a completely lethal method of data gathering and should not be performed on species from small or declining populations. Because they may provide information about a trait(s) under selection, they are not neutral markers and should not be used for population-level studies. On the positive side, allozymes are codominant markers (we can see both the homozygote and heterozygote individuals), meaning patterns of Mendelian inheritance can be observed from this data type. A literature search should be performed to evaluate any published allozyme data as there are data for several endangered species including the Karner blue butterfly (*Lycaeides melissa samuelis*) (Packer et al. 1998) and the Scarce Heath butterfly (*Coenonympha hero*), a species endangered in Europe (Cassel and Tammaru 2003).

RAPDs

Random amplified polymorphic DNA (RAPD) are an example of a dominant molecular marker, meaning they only show the presence of a dominant allele and do not distinguish between homozygotes and heterozygotes. A recent evaluation of molecular methods in endangered Lepidoptera showed that RAPDs, while requiring little DNA to run, are not highly informative at the population level, are not reproducible, and are therefore not as valuable as other molecular markers for conservation questions (Turlure et al. 2014).

Sequence Data: Barcoding, Mitochondrial Genes, and Beyond

With the advent of Sanger sequencing in the late 1970s, more practitioners have turned to DNA sequence data to answer genetic questions. Sequence data have proven very informative for barcoding, which compares sequences from a commonly sequenced mitochondrial gene such as cytochrome oxidase subunit 1 (CO1) to publicly-available databases like the National Center for Biotechnology Information (NCBI). Barcoding may provide identification of new species sequences (www.lepbarcoding.org) and provide information about species boundaries. There is a great utility to barcoding with standard sets of genes, but barcoding is also controversial, especially when it is used to delimit endangered species boundaries. Rates of evolution for mitochondrial gene mutation do not match evolution in other parts of the genome and may provide evidence that is discordant with other molecular markers, morphological data, or biological relevance (Wiemers and Fiedler 2007; Forister et al. 2008). Data from mitochondrial gene sequencing has been effectively used in considering several of the butterflies in the case studies; notably the Karner blue butterfly in issues of mitochondrial introgression from the Melissa blue butterfly (Gompert et al. 2006) and the relationships between newly discovered populations of the Mitchell's satyr butterfly and the two federally-protected subspecies (Hamm et al. 2014a).

Next-generation sequencing technologies (e.g. 454 ®, Illumina®, Ion Torrent ®, PacBio ®, etc.) are new sequencing platforms that generate millions of base pairs of DNA sequence data. We use bioinformatics methods to search the large data sets for both coding and non-coding regions of the genomes and these technologies have provided a great base for genetic work in non-model organisms (Saarinen and Austin 2010). Gompert et al. (2010) performed a pyrosequencing run (using 454® platform) and generated hundreds of thousands of reads to evaluate the phylogeographic history of the Karner blue butterfly genus, *Lycaeides*. A comprehensive evaluation of 101 members of the Monarch butterfly genus (*Danaus*) was recently completed using Illumina sequencing. This project identified 32 million SNPs (single nucleotide polymorphisms) to characterize both migratory patterns and warning coloration (Zhan et al. 2014). These new sequencing technologies have also provided valuable information about microsatellites, a marker of great utility to conservation.

Microsatellites

Microsatellites, also known as simple sequence repeats (SSTs) or short tandem repeats (STRs) are codominant markers that allow for the characterization of both heterozygotes and homozygotes. Microsatellites have been the molecular marker of choice for at least the last 10 years (Frankham et al. 2003) as they provide an excellent resolution and characterization of genetic diversity. Additionally, they can be amplified using small quantities of DNA that can be gathered from non-lethal tissue collection (Crawford et al. 2013; Hamm et al. 2010). Microsatellites are also highly variable markers (usually with multiple alleles per locus). They provide a detectable signal faster than a marker (like a mitochondrial gene) that would mutate more slowly. By using many microsatellite loci located across the genome, we are able to produce a better, more complete picture of neutral genetic diversity from the individuals sampled. Because microsatellites are so variable, they are also usually species-specific, which reduces the probability of cross-contamination between samples of different species.

Several of the disadvantages to using microsatellites in endangered species work have recently been alleviated with the advent of new next – generation sequencing technologies (Saarinen and Austin 2010). These technologies include sequencing longer reads of the genome and from these longer reads we are able to use bioinformatics methods to identify microsatellite loci. This prevents the need for tedious, expensive, often unsuccessful laboratory work that had been traditionally used to identify microsatellites. New sequencing technologies (including 454® and other platforms) are also becoming less expensive and new methods have been developed that allow for the sequencing of multiple species in single runs. One study (Jennings et al. 2011) suggests that all U.S. federally-endangered species may be sequenced and microsatellites developed for them for <$500,000 USD.

Microsatellites in Lepidoptera, while certainly useful, are not without their own issues (Meglécz et al. 2004; Zhang 2004). Null alleles are a frequently noted problem in which primers fail to amplify the targeted microsatellite region (due to base pair changes at the priming sites). This may lead to inaccurate characterization at the population-level (i.e. actual heterozygotes may be counted as homozygotes if the second allele drops out due to a null allele problem) as well as a dataset with lots of missing data. However, awareness of these problems allows us to perform additional evaluations of the data in order to better interpret the results. For example, free programs (e.g. Micro-Checker) (Van Oosterhout et al. 2004) can be used to estimate the number of null alleles, large alleles that dropped out, and problems of stutter peaks in a population to inform which loci should be used for larger-scale testing and interpretation.

Screening for Pathogens

Several pathogens that infect butterfly species have been demonstrated to have negative consequences on the butterfly population. Considerable research has focused on the α-proteobacterium *Wolbachia* and Hamm et al. (2014b) showed that 19 of 22 species of Lepidoptera of conservation concern in the U.S. were infected. This rickettsial symbiont has been found in over 65 % of arthropod species and is known to be maternally-inherited (Werren 1997). *Wolbachia* infection may have several consequences on its host phenotype by: (1) turning male embryos to female embryos; (2) killing all male embryos; (3) producing cytoplasmic incompatibility between males and females. The type of phenotype induced in the host depends on the strain(s) of infecting *Wolbachia* as well as the host identity. Cytoplasmic incompatibility describes the situation where adult males infected with *Wolbachia* are unable to reproduce with uninfected adult females (or females infected with a different strain). Because of the negative phenotypes produced by this pathogen, *Wolbachia* presence needs to be closely evaluated in butterfly species of special concern where management plans consider either translocation or captive propagation. Failure to evaluate populations for *Wolbachia* (and specific determination of the strain type) may result in negative consequences (all female populations) that are the result of infection by one or multiple strains.

Basic genetic methods (PCR analyses) can be performed to positively identify the presence of *Wolbachia* and strain identification can be verified by DNA sequence analysis. I recommend testing multiple individuals in a population to evaluate the level of infection in a population. Additionally, populations at all parts of the species range need to be considered, as it has been shown that not all parts of a region may be infected, and that infection status can change over time (Nice et al. 2009). There are some correlations between strain identity and the phenotype induced in the host (Zhou et al. 1998; Baldo et al. 2006). However, phenotype cannot be inferred from strain type because different strains may behave differently in different host species and it is also possible for multiple strain infection to produce alternate phenotypes. Only experimental laboratory crosses can definitively describe the impact of strain types on host species (Hiroki et al. 2004), a procedure that is likely prohibitive for most endangered taxa.

While *Wolbachia* has received the majority of the research-focus, other pathogens have also been evaluated in the Lepidoptera and should be considered (and screened for as possible) when translocation and captive breeding are proposed as part of a management plan. Duron et al. (2008) describe *Cardinium*, *Arsenophonus*, and *Spiroplasma* as additional causes for concern. Phenotypes presented from these infections also include male-killing, and have been documented in other Lepidopteran species (Jiggins et al. 1998, 2000; Dyson et al. 2002).

Lastly, baculovirus infections are important to consider in a captive propagation program. Baculoviruses have been used to control populations of Lepidopteran pest populations (Moscardi 1999), and their presence may wreak havoc by causing a "wasting disease" in captive-bred larvae (personal observation). Because these viruses spread so quickly, molecular methods may only provide an after-the-fact confirmation of the virus that destroyed a captive colony.

Statistical Considerations

Once the decision to perform a genetics study on an endangered species has been made, it is important to consider statistical details. If you are performing a population-level study (e.g. using microsatellite markers), it is important to sample at least 30 individuals per population. Statisticians usually recommend sampling 31–32 individuals from each population if possible. Thirty individuals usually represent the minimum number of individuals needed to accurately capture the genetic diversity that remains in a population. Too few individuals do not provide enough genetic information and therefore are not useful for decision making. Therefore, if you cannot sample 30 individuals, I do not recommend that you perform a population-level study. For univoltine species, it may be possible to sample 30 individuals over a 2 year period. This will still provide a relatively robust estimate of population-level diversity, even if the samples represent two generations.

Because we are working with endangered/rare/threatened species, it is safe to assume that they have already lost some level of genetic diversity. Therefore, it is important to use multiple molecular markers (covering a greater portion of the genome) to get a strong statistical signal useful for analyses of genetic diversity, gene flow between groups, population differentiation, etc. Projects using microsatellites should therefore use multiple markers. There are many studies with non-endangered species that use <10 microsatellite markers (e.g. Caplins et al. 2014) that are able to recover a meaningful statistical signal due to the diversity (i.e. number of alleles) present at those microsatellite markers. Because the microsatellite markers for endangered species may not be as information rich (i.e. may not have as many alleles per locus), I recommend 20+ microsatellite markers. Consultation with a statistician is helpful in evaluating how many markers are necessary to detect genetic signals. A power analysis can be performed in different statistical software packages to help determine the number of microsatellite numbers needed (Ryman et al. 2006).

For studies using mitochondrial markers for sequence data, it is important to make sure that all extant populations are sampled. Typically one or two individuals per population are all that are needed. Studies evaluating pathogen infection need to sample a predetermined percentage of the population, as the fraction of a population that is infected will need to be determined. Multiple individuals will need to be sampled to determine whether multiple strains of a pathogen are present in the population.

Implementation: How Do I Carry Out an Effective Conservation Genetics Project?

I have seen projects succeed and fail in several ways, and I provide this as a chronological overview to assist in planning and help increase the success of a project. Oftentimes a focus on the "genetic" component of the project is emphasized and this may to the detriment of other areas. Other projects have suffered due to lapses in communication between field researchers and land managers. Expectations between involved parties should be articulated at each step of the planning process. In many ways the "genetic" components are the most-straight-forward; experiments work or they don't and alternative methods can be explored. Here, I present a series of recommended steps (Fig. 1).

Before the Field Season Begins: Planning and Communication

Project Planning The value of project planning cannot be underscored enough. The levels of planning are varied (both in terms of people involved and physical locations) and this step should take place far in advance of the field season. In some instances, permits may take up to 1 year to acquire. Any project that involves netting or handling a butterfly and taking a wing fragment or leg sample from a state or federally endangered species requires a permit.

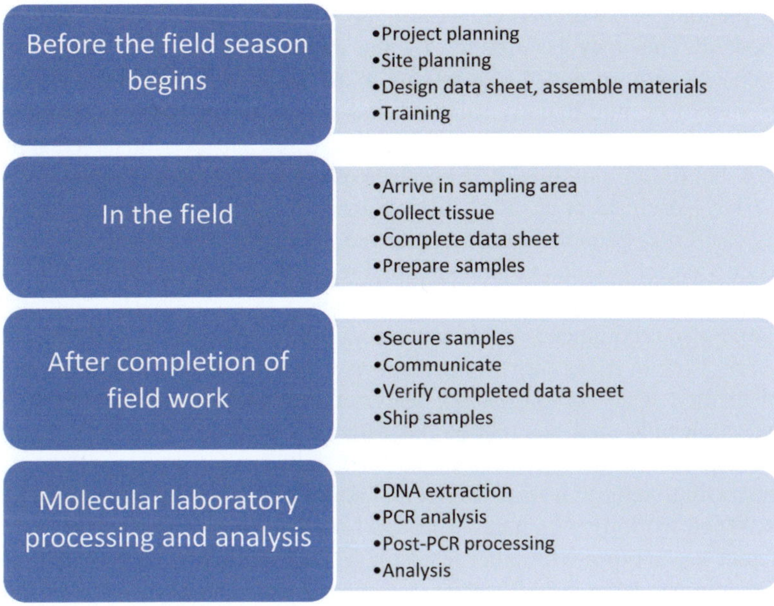

Fig. 1 Proposed framework for conducting conservation genetics work for endangered butterfly species

Due to time constraints, I recommend a strong emphasis on initial planning and communicating. Conference calls to coordinate with stakeholders are an essential first step. I also recommend developing a time line for deliverables. Deliverables in this instance can mean field data sheets as well as raw samples and analyzed molecular data. When stakeholders have a timeline for expectations, they are better equipped to deal with adversity and account for the actual time to complete tasks. Planning before the collection of samples also needs to be coordinated with the molecular laboratory contracted to perform the work. Is the lab aware of the timeline that samples will be collected? Often times, samples will be processed by a graduate student or as part of a project. Communication with the laboratory investigator is important; will personnel be available to receive and process samples? Can the student complete the work in the necessary timeline? Who is paying for different parts of the project? Is there a contingency plan if the student leaves the laboratory? Will another researcher finish the project? Working with an academic institution is usually a mutually-beneficial situation for conservation genetics projects, and they typically offer labor cost savings as well. Projects are best executed with communication and transparency. I also advocate for some discussion amongst stakeholders for other contingency plans (regarding weather, personnel, permit issues). Plans do not to be extensively detailed, but points of contact for different issues should be identified.

The planning or working group also needs to decide on the tissue type to be collected. This may be affected by the endangered status of the butterflies, the taxonomic group, and the questions to be addressed. Certainly, endangered species should be non-lethally sampled and extreme care taken to avoid unduly harming individuals. There have been several research papers published that provide evidence that tissue sampling does not decrease fitness in butterfly species (Hamm et al. 2010; Koscinski et al. 2011; Crawford et al. 2013). For population genetic studies, small non-gametic tissue is preferred. I sample wing tissue in Lycaenidae (it is usually easier to collect than individual legs which may be too small to sample in the field). In larger species (e.g. many members of Papilionidae, Nymphalidae, Pieridae), I also recommend wing fragment sampling as opposed to legs. I perceive removal of a leg in many larger butterflies as invasive due to the amount of tissue present in their legs. As long as a portion of the hemolymph-containing vein is sampled, molecular analyses will be possible with wing tissue and not legs. Other families (e.g. Hesperidae) are very active when netted and it is not possible to easily hold them still to remove a wing fragment. I advocate for single leg removal in these groups. When working with a new species, I recommend discussing pros and cons of different tissue types with other research groups. Practicing on surrogate species is also recommended and will ensure that field workers encounter fewer problems in the field when handling the endangered taxon.

Site Planning Early planning stages should include conversations of the lands to be surveyed. You must also secure proper permits (for different lands, for threatened species, etc.). As you develop your sampling plan, you need to define specific collecting areas. Generate maps that can be given to field workers to ensure that they can properly find the field site and are sampling at the correct location. This is especially useful if transect counts are to be performed along with tissue sampling for genetic work. I have found that transect counting and tissue sampling can be done by the same individual(s) and will provide an excellent comparison between census population estimates and genetic population estimates (Saarinen et al. 2010). Proper planning will ensure the species are sampled across their range, which is very useful for evaluating gene flow, genetic differentiation, and levels of pathogen infection.

Design Data Sheet and Assemble Materials To ensure that the proper types of data are collected, I recommend communication with end-users on data sheet design. I have provided an example of a field data sheet that can be used in a sampling project (Fig. 2). Data sheets need to be easy to read and use for field scientists and volunteers as well as those summarizing and interpreting their data. The types of field data to be collected relating to abiotic conditions (weather, precipitation, soil conditions, percent cloud cover, temperature) are important values that may explain butterfly abundances.

I also recommend collecting basic data on other biota of importance to butterflies, such as the identity and abundance of larval host plants, species of flowering plants, percent of field flowering, etc. These types of data are important for tracking

Site name____________ Plant descriptions (larval host plants, available nectar and types)____________

Date______ Time______ Personnel names and contact info____________

Weather: %cloud cover______ wind______ humidity______ temp______ Other notes:____________

Species	ID#	Sex	Location description	Time	GPS	Photos?	Collector	Notes

Fig. 2 A sample datasheet for field work and tissue collection

phenologies and potential phenological mismatches as we continue further into the era of global and regional climate change. It is difficult to collect all of these data without designing the sheet to be too cumbersome. However, these are valuable pieces of data that can only be collected on the ground. Many of us have been in planning sessions and heard the wish to go back in time to collect data on flowering abundances and timing.

The data sheet/package should also have a list of abbreviations to be used. This will avoid confusion for the data organizer. The sheet will also explain the unique identifiers for individual species and tissue collections. I recommend the first two letters of genus and species (*Oarisma poweshiek* = Oapo1, Oapo2, Oapo3, etc.). Numbers should never be duplicated for a single species. Initial coordination will be helpful in ensuring that the molecular laboratory receiving samples from several field locations does not end up with multiple samples. Many times I have received boxes of samples, all beginning with "Sample #1." This leads to confusion for the molecular laboratory. At best, they have to spend considerable time relabeling and matching samples with field notes and, at worst, there is potential for incorrectly labeling specimens or mismatching sample vials with their collection information. To combat this problem, I recommend that the molecular laboratory handling the tissue to mail out materials before the field season begins. This way, tissue collection vials can be pre-labeled and will ensure that duplicate numbers are not returned. It is important that people sampling tissue in the field match their datasheets to the pre-numbered vials.

In addition to the numbered vials with ethanol to be used for collection and storage of tissue samples, a variety of other materials need to be assembled to prepare for field work. Vials should be kept in a cool dark place as much as possible; I recommend a small cooler with a few ice packs. Numbered vials should be in a hard-sided cardboard or plastic box. Additional materials to be included in a field bag for the specific purpose of tissue collection and processing:

- Aerial butterfly net
- Squirt bottle of 70+ % ethanol

- Box of Kimwipes (Kimberley Clark ®)
- Several types of forceps (philatelic forceps with rounded and square ends)
- Sharpie ® markers (fine point, not ultra-fine point). Black usually works well unless the species is dark, then I recommend a metallic Sharpie
- Clip board to hold data sheet and act as a solid surface for writing
- Tupperware or bag to hold used Kimwipes
- Camera

Additional communication between those doing the field work, data sheet compiling, and the molecular analyses also need to make a decision on method of data communication (e.g. mailing hard copies of sheets, emailing sheets, uploading to website). Information sharing should happen before the field season begins in order to prevent delays later. Information to be shared includes names, phone numbers, email addresses, physical mailing addresses (and limitations with these), as well as accessibility and permissions to upload data to web-based data management systems.

Training All field-based team members need to be equipped with field guides (for both plant and butterfly identification) and maps to sampling sites. Collectors' identification abilities may need to be verified before the field season begins. This is especially important in cases where endangered species may be confused with common species. I encourage photographs to be taken of butterflies where a clear identification is difficult. Evidence of photographs and comments should be clear on the data sheet. Detailed note taking is important and will prevent undertaking expensive and time-consuming molecular work on a species that is not the focus of the study.

Collectors must also be trained in proper butterfly handling (see section "Collect tissue" for details). I recommend practicing with a surrogate species of the same butterfly family. Training should include a discussion of problem-solving in the field and contingency plans. Training must also ensure that all field collectors are aware of the problem of cross-contamination between samples. *Failure to properly clean equipment between samples will lead to project failure.*

Lastly, if a mark-release-recapture study (MRR) will accompany the field work, all field workers need to use the same protocols and naming conventions and the field data sheet should be modified to allow for easy inclusion of MRR data.

In the Field: Preparation and Attention to Detail

Arrive in Sampling Area Upon arrival to the field site, it is important to begin careful note taking. Time spent filling out field notes forms will yield valuable data later. I recommend taking some time to note general conditions and to record specific observation, with quantitative (i.e. numerical) measurements when possible. Take time to photograph the field sampling area to support the field notes and provide additional evidence. If it is part of the original pan, complete initial formal transect assessments. After initial surveying to evaluate the quality of the site, plants, and compile a preliminary species list of butterflies, it is time to prepare for the target species tissue collection. Take a moment to arrange materials and vials for tissue collection. Most field sites are too large to set up a "base camp" and as such all your materials will be with you in your backpack. Make sure you can easily swing your net and chase species with your backpack on.

Collect Tissue Locate and gently net the target species of butterfly. Flip the net over so that the butterfly will not escape. If the butterfly is caught on the ground, hold the end of the net up so that the butterfly will fly up into the tip of the net. It is easier to work with a butterfly if it is confined to a small part of the net. Record time caught and sex (if known) with unique identifier (ID). This ID should be on the sample vial (make sure that you write down the correct, corresponding unique ID on the datasheet). Set up forceps, Sharpie markers, Kimwipes and ethanol vials so that all are accessible. Ensure that vials are secure and will not spill. Prepare (unscrew) only the vial to hold the next tissue sample.

Gently immobilize the butterfly by holding it between thumb and forefinger at its thorax. Take care not to touch its wings unnecessarily. Place a dot on the ventral side of the butterfly wing with a Sharpie (this may be done through the net or once the butterfly is manually removed from the net). If a mark-release-recapture study is done, mark the ID number on the wing. Keep Sharpie capped when not in use. Fresh Sharpies leave the best marks on butterfly wings.

Leg Removal While still holding the butterfly at the thorax between the thumb and forefinger, remove a fore or mid leg with fine-tipped forceps. This may also be done through the butterfly net itself (i.e. butterfly was immobilized through the net but not removed from the net). If the gauge of the net is too small, remove the butterfly from the net and then remove the leg. I typically advocate for removal of a fore or mid leg, as opposed to a hind leg which may be important in providing stability to the butterfly.

Wing Fragment Removal For wing tissue removal, remove the butterfly from the net. Use thumb and forefinger or flat forceps to hold the butterfly still. Use a second set of flat forceps to remove a small piece of wing tissue from the hind wing (Fig. 3). The size of the fragment to be removed need only be 2 mm^2, however it is essential to include a small portion of a wing vein in the tissue collection. As long as the wings are not crushed or ripped from the thorax, this method of sampling yields adequate DNA and is non-lethal to the butterfly. Touching the butterfly with wet fingers (from ethanol) is detrimental to a butterfly in the field. Take care not to apply too much pressure to the forceps as it may crush the wings veins. Careless removal of tissue that tugs at the point of attachment of the wing to the thorax is also potentially lethal to the butterfly. I advocate for *extreme care* in removing the wing tissue and note that I have recovered DNA from a variety of wing veins (e.g. anal, cubital, and medial). The anal and cubital veins are preferred because they are often easier to sample (especially the A3 vein). Wing veins contain hemolymph

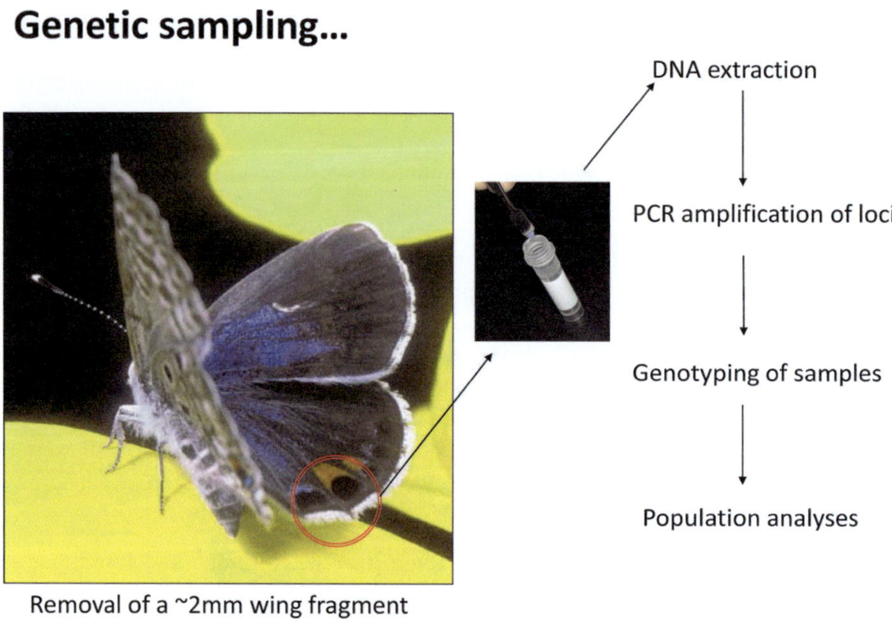

Fig. 3 Wing fragment genetic sampling technique. Shown here on *Cyclargus thomasi bethunebakeri*, the endangered Miami blue butterfly (Photography by Jaret Daniels, McGuire Center for Lepidoptera and Biodiversity, Florida Museum of Natural History)

(i.e. blood) and hemolymph has DNA, therefore the yield of DNA will be greatly increased with the inclusion of a portion of a vein. DNA may still be isolated from a wing fragment that lacks a vein, but it is of very limited utility because of the very low concentration of DNA that can be recovered. There are very small bits of tissue at the point of attachment between a wing scale and the wing surface, and this leads to minimal recovery of DNA. Therefore, to increase actual genetic utility, a portion of a wing vein must be included (Hamm et al. 2010).

As you place the tissue (leg or wing fragment) in its labeled, open vial, ensure the tissue is submerged in the ethanol. Use 95–100 % ethanol in the storage vials. Securely cap the vial and return it to the storage box. When not in use, keep the box of collected tissues in a cool, dark place (preferably a small cooler that has been brought into the field for this purpose).

Take care when releasing the butterfly back into the field. I recommend releasing the butterfly on a nectar source, preferably in a shady area. Because the butterfly has just been handled, it may be a bit slow to recover and need to recover in a safe location. Be assured that butterflies may also recover very quickly after minimal handling. However, I have anecdotal evidence of recently released butterflies being grabbed by foraging dragonflies. As such, release the butterfly in an area where it can safely recover and not be predated or dry out in the hot sun.

One of the most important parts of tissue collection is the proper cleaning of materials between sample collections. You must clean ALL MATERIALS (e.g. all forceps used, your own fingers, etc.) that contacted the butterfly. Use ethanol in the squirt bottle and Kimwipes to remove scales from forceps. Dry all materials well with additional Kimwipes before sampling another individual. Make sure to always clean forceps at the beginning and end of a sampling day as well. Upon return from the field, wash forceps with 10 % bleach solution, rinse very well, and dry and store them before conducting additional sampling. Because the molecular work occurs on very small amounts of tissue; even individual scales from one specimen in another specimen's vial act as contamination and can lead to misleading genetic results. Mixtures of scales in a vial can also lead to potentially elevated levels of genetic diversity and therefore inaccurate results. We have found that immobilizing butterflies in glassine envelopes to perform tissue sampling may actually lead to more air-borne scales and potential contamination. All techniques that minimize the amount of scales in the collecting environment must be enforced. This may limit the amount of handling and/or photography that occur with tissue collection, but it is imperative to minimize contamination between samples.

Complete Data Sheet Ensure that all field notes are properly completed. Note time collection began and ended, field notes, weather, % cloud cover, wind, temperature, other species of butterflies seen. Circle or highlight important findings or aberrations for future investigation or so that you may relay findings to the rest of the management team.

After Completion of Field Work: Organization and Timeliness

Secure Samples Ensure that labeled samples are all in labeled boxes and not left in field bags. Samples can be in a refrigerator or cooler indoors as they await shipping. Do not freeze samples unless previously instructed by the lead researcher.

Communicate Notify the lead researcher and other team members of your field experiences. This may help other field collectors/members of your team. Quick email correspondences work best during the busy field season.

Verify Completed Data Sheet Memories fade, make sure your descriptions of the field site and notes are complete! Upload field notes to project website or coordinator if possible.

Ship Samples Verify address and method of shipping. Notify contact person that samples are being mailed. Take care not to mail on a weekend or holiday (when samples may be stranded in a hot location). Because you are mailing liquid in the U.S. mail, all vials must be double bagged. Tape sample boxes closed, and place them in a sealed, gallon Ziploc bag and then place that in another sealed Ziploc bag. Additional padding material may need to be added to the shipping box. You will also need to add enough absorbent material to soak up any ethanol that leaks from the vials. It is legal to mail tissue in ethanol by FedEx, UPS, and DHL as long as the total volume does not exceed 30 mL. The small volume of ethanol present in each vial is exempt from regulation (DOT 173.4 and USPS Publication 52 (334)). The outer shipping box should be labeled "This package conforms to 49 CFR 173.4."

Molecular Laboratory Processing and Analysis

The following four steps will be carried out in a dedicated molecular laboratory (e.g. an academic laboratory or government facility). I include this section in an effort to let the tissue collectors know what happens to the samples after they mail them. Molecular work should not exist "behind the curtain" and I want to allow interested parties and stakeholders some insight into the procedures of the molecular laboratory.

DNA Extraction Tissue samples will be individually processed in order to extract the genomic DNA (DNA from the mitochondria and nucleus together) from each sample. This process typically takes 2 days. Samples are individually homogenized in the presence of a cell lysis buffer and an additional enzyme that breaks down the cell membrane. Samples are incubated overnight to allow for cells to open up and DNA to enter the buffer solution. On the second day, procedures are performed to remove contaminants from the solution (proteins, pigments, etc.) and lead to an isolation of concentrated DNA for further use. At the end of the extraction period, it is important to save multiple aliquots of DNA extract. This will prevent the freeze-thaw cycle from degrading the DNA template as it is used for subsequent processing (e.g. amplifying for multiple microsatellite markers, identification of *Wolbachia*, etc.). Saving multiple aliquots will also allow them to be shared with different laboratories. Whenever possible, DNA should be stored in multiple freezers to protect against total loss due to power surges, crashes, or freezer breakdown.

If extracted DNA is of very low yield (ng/uL), it is possibly to amplify the DNA template with commercially available kits. I recommend REPLI-g minikit (Qiagen®, Valenica, CA) to increase the volume of DNA. This kit uses a rolling circle enzyme to increase the overall yield of DNA. This process cannot improve the quality of the DNA template (if it is very old or degraded DNA it cannot repair damage). It also amplifies any contamination (e.g. bacterial) present in the DNA sample. I have used this kit to essentially "make more" DNA when template stocks have gotten low. The results from subsequent PCR analyses have matched between the original DNA extract and the DNA that had been amplified via REPLI-g (Saarinen, personal communication).

PCR Analysis Genomic DNA is used to perform PCR (polymerase chain reaction) of selected molecular markers by isolating and amplifying a region of interest using DNA polymerase and additional nucleotides. Specific primers are added to the reaction mixture to amplify regions of the DNA that code for a gene of interest or a highly variable region (e.g. a microsatellite). Specific primers can also be used to evaluate a sample for pathogen contamination and strain type (e.g. *Wolbachia*). A PCR reaction may be viewed on an agarose gel.

Post-PCR Processing Some molecular results may be directly read from an agarose gel (i.e. presence of absence of a pathogen). However other results require further instrumentation for processing (e.g. evaluation of sequence data and microsatellite genotyping). Therefore, the lab will either conduct this step itself or send it to a genetics core facility for final processing. Major research centers and academic institutions have the equipment to perform these analyses and then will email the lead researcher the completed results. I recommend that special consideration be given to the facility that will genotype microsatellite data. Because this type of data is based on read length (e.g. 180, 183 bp) and not the identity of the nucleotides (e.g. AT, CG), there may be slight differences in read length interpretation between sequencing instruments. Even the same model of a machine at two different institutions may provide slightly different results. Therefore, I recommend conducting a microsatellite genotyping project on a single machine. If this is not possible (i.e. machine breaks, becomes unavailable) there are calibrating techniques to unify the datasets produced by different machines.

Analysis Raw molecular data from the sequencing platform will always need to be processed before it can be imported into statistical programs. This may require specific, proprietary software (e.g. GeneMarker®, GeneMapper®). Examples of sample output are shown in Fig. 4. Discuss this with the genetics core facility to ensure that this initial data translation can be accomplished. Once this step is complete, I recommend a general conversion program to help organize the

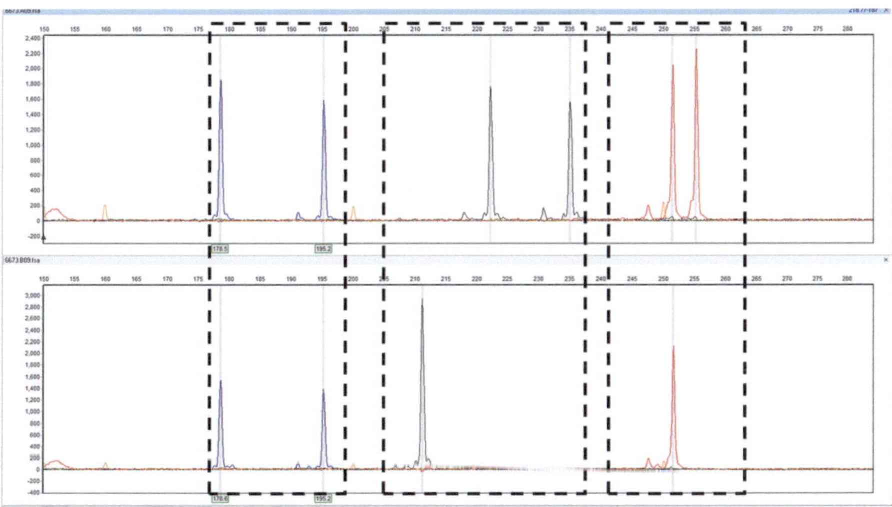

Fig. 4 Example of microsatellite genotyping data for two *Lycaeides melissa samuelis* (Karner blue butterfly) individuals. *Dashed lines box* out three unique microsatellite loci. In locus one (*left-most box*), both individuals are heterozygous and have alleles of the same size. Locus two (*middle box*) shows individual 1 as a heterozygote and individual 2 as a homozygote (and with a different allele). Locus three (*right-most box*) shows individual 1 as a heterozygote and individual 2 as a homozygote, and with an allele in common with individual 1

molecular outputs. A program (e.g. Convert 1.31, available for free download: http://www.agriculture.purdue.edu/fnr/html/faculty/rhodes/students%20and%20staff/glaubitz/software.htm) will allow for your data to be exported as different file types that are needed for specific analyses. I highly recommend using software to perform these tasks as a way to minimize error. Once properly formatted, genetic analyses can begin to answer questions of genetic diversity, gene flow, inbreeding, identify unique genetic diversity, etc. Some of the most user-friendly and commonly used analysis programs are: GenAlEx 6.5, Genetic Analysis in Excel (Peakall and Smouse 2012), Arlequin 3.5 (Excoffier and Lischer 2010, Structure 2.3.4 (Pritchard et al. 2000), Geneious (Kearse et al. 2012). An overview of population genetic software packages, phylogenetic packages, and free downloads is available at: https://courses.washington.edu/fish543/Software.htm.

Case Studies and Outcomes

There is a wealth of information on the genetics of butterflies (e.g. phylogenetics, eyespot patterns, mimicry complexes, etc.) but few studies directly connect the genetics of endangered butterflies to conservation programs, specifically for planning and management purposes. The inclusion of genetic data has become profoundly important to both the listing status for the definition of populations and species boundaries (Morin et al. 2010; Laikre et al. 2009). Under the US Endangered Species Act, invertebrates (including butterflies) are provided with full species-level protection for subspecies taxon designations. Species and subspecies-level designations (as results of gene flow and unique evolutionary lineages) are therefore important for listing status as well as for meeting management thresholds and delisting criteria. Additionally, population-level diversity measures are important as a goal of management is to conserve remaining genetic diversity. In this section, I highlight three case studies of federally-endangered butterflies in the United States. Each of these studies has made the connection between molecular research and conservation/management planning.

*Karner Blue (*Lycaeides melissa samuelis, *Lycaenidae)*

The Karner blue butterfly represents "a model endangered butterfly" in the United States as it has been the subject of multiple research projects and programs. Listed as federally-endangered in 1992, this taxon has a long history in the conservation world. The Karner blue was originally described by the novelist Vladimir Nabokov (1949) and issues of taxonomic uncertainty have followed the taxon for decades (Nabokov 1975). Taxonomic discrepancies stemmed from its subspecies designation and conflicting accounts related to the uniqueness of its evolutionary lineage when compared to the nominate species, *Lycaeides melissa melissa*. Genetic work of 34 allozymes indicated that *L. m. samuelis* was not differentiated from *L. m. melissa* (Packer et al. 1998). Less than a decade later, molecular methods had advanced and new methods were used to evaluate the species status of the *Lycaeides* group. This work evaluated not only the two members of the *L. melissa* complex, but also the sister species. *L. idas* (whose taxonomic certainty was not in question). These studies showed that mitochondrial introgression had occurred from *L. melissa melissa* to *L. melissa samuelis* in the past, but that both are likely unique genetic lineages (Nice et al. 2005; Gompert et al. 2006). Gompert et al. (2006) performed a useful evaluation of the federally-endangered Karner blue butterfly haplotypes using sequence data and AFLPs. Most recently, a full genomic scan of the *L. m. samuelis* genome (using 454 Roche ® next-generation sequencing technology) has provided additional, definitive evidence that all three butterfly taxa are indeed distinct evolutionary lineages (Forister et al. 2011). In this extensive study Forister et al. (2011) sequenced multiple individuals from populations across the ranges of each of the three taxa. Extensive sequencing and modeling work showed the three taxa were unique and the study provided estimates of similar (and low) levels of gene flow between *L. m. melissa* and *L. idas* as from *L. m. melissa* and *L. m. samuelis*. This study provides statistical support to the argument that the Karner blue is deserving of its own specific status, *L. samuelis*, an opinion that Vladimir Nabokov himself put forth in the mid-twentieth century. Using current genetic techniques, several authors have successfully unraveled the taxonomic quandary of this species.

While this information did not change endangered species listing status per se, it did help provide further insight into the taxon, as well as add support to management because of hybridization with a sister species (*L. m. melissa*) which has putatively led to *Wolbachia* infection in the endangered Karner blue. Nice et al. (2009) screened populations of the Karner blue range-wide for infection with *Wolbachia*. They discovered high infection rates in the western populations of the Karner. Sequencing of the *Wolbachia* strain show similarity with the strain infecting the Melissa blue, in the area where the butterflies co-occur and may hybridize. Additional evidence of mitochondrial introgression from the Melissa blue to the Karner provide support for the hypothesis that *Wolbachia* infection originated from the Melissa blue. Both mitochondria and *Wolbachia* are vertically-transmitted. Only one individual from an eastern Karner blue population was infected and this was from a different strain of *Wolbachia*. Serious concerns about how *Wolbachia* infection could further depress already small populations were raised as a result of these findings. Great care and consideration in management activities was therefore enacted so as not to infect the eastern populations with the western strain. Knowledge of *Wolbachia* infection in the different population groups was especially useful as it prevented the spread of the pathogen. Identification of the source of the *Wolbachia* (from the sister species, Melissa blue) was useful and provided evidence to minimize contact between the groups for conservation purposes.

A range-wide population genetics assessment of the Karner blue is nearing completion in 2014–2015. This project gathered wing clip samples from 30 individuals per population over the entire range of the species. The genetics portion was led by Dr. Tim King (USGS, Leetown Science Center) in conjunction with Dr. Jessica Hellmann and researchers at the University of Notre Dame. A set of polymorphic microsatellite markers was developed to evaluate genetic diversity across the species range as well as to determine the level of gene flow (based on a measure of genetic similarity) between populations. As part of a similar but more localized study, I am also using these microsatellite markers to evaluate population dynamics and dispersal at the metapopulation level in a population of Karner blue butterflies in Michigan. Results from the large and small-scale population dynamics will provide much needed insight into the neutral genetic diversity and patterns of gene flow, both of which may be useful for management planning.

*Mitchell's Satyr (*Neonympha mitchelli*, Satyridae)*

Neonympha mitchelli, at the time of this publication in 2015, has two unique subspecies that are listed as federally-endangered (*N. m. mitchelli* and *N. m. francisi*). Extant populations of Mitchell's satyr (*N. m. mitchelli*) are found in Michigan, Indiana, North Carolina, Alabama, Mississippi, and Virginia. The St. Francis satyr (*N. m. francisi*) is only known from a single site in North Carolina, approximately 200 km away from the nearest *N. m. mitchellii* (Hamm et al. 2014a). Genetic analysis provides strong evidence for the support that *N. mitchelli* is monophyletic (one evolutionary lineage) and that its two described subspecies are distinct and based on geographic location (Hamm et al. 2014a). Newly discovered populations of *N. mitchelli* were found in the extreme southern portion of the species range in Alabama and northeastern Mississippi, but a clear relationship with the other subspecies is unclear despite considerable genetic testing of six markers (five nuclear genes and one mitochondrial gene) (Hamm et al. 2014a). I agree with the author's recommendation that all three populations should be managed separately, as they all represent unique genetic diversity. In that study, genetic data were crucial to providing evidence that the newly discovered southern populations should not be included as populations of either of the other subspecies (Hamm 2012). If they did have sufficient genetic similarity to one of the other subspecies, they could be used as additional populations that would provide support for their de-listing by demonstrating that more populations existed (as proposed in the management plan).

Recently, Hamm et al. (2014b) screened 22 species of conservation concern, including *N.m. mitchelli*, for *Wolbachia*. This project showed that 19 of the 22 species tested positive for *Wolbachia* and the authors propose follow up work to evaluate how pervasive the bacteria is across the various species' ranges. The discovery of *Wolbachia* in *N. m. mitchelli* will undoubtedly be considered in future management plans for the taxon.

*Miami Blue (*Cyclargus thomasi bethunebakeri*, Lycaenidae)*

The Miami blue butterfly was listed as a federally-endangered butterfly in 2012 and its precipitous decline has been tracked for several decades (Calhoun et al. 2002; Daniels et al. 2008). This once common south Florida butterfly now exists as a small, island metapopulation in the extreme southern portion of the state. Concerns about remaining genetic diversity and gene flow between habitat patches and metapopulation dynamics prompted genetic study (Saarinen 2009). Traditional molecular laboratory methods were used to develop 12 polymorphic microsatellite loci (Saarinen et al. 2009). These markers were used to evaluate gene flow and genetic diversity in a variety of samples, including the metapopulation of south Florida, museum collections of historical populations, and the captive colony of butterflies housed at the University of Florida.

Genetic research using microsatellite markers confirmed the metapopulation structure of this taxon in the wild, and that high levels of gene flow connected patches of suitable habitat (Saarinen et al. 2014). There was no evidence of a genetic bottleneck in one metapopulation, and there was minor evidence for a bottleneck in another (indicating the area may have been recently colonized or had undergone a genetic bottleneck due to a major loss of individuals and population crash). Evaluation of the captive colony at the University of Florida did show evidence of a genetic bottleneck (Saarinen et al. 2009). These results provided evidence to support that a captive colony should regularly be infused with new genetic material from the wild to prevent bottlenecks and loss of diversity. A rescue effect was noted in the captive colony when new stock was incorporated from the wild (Saarinen 2009).

Additional analyses show that field census estimates are consistent with the effective population size (Ne) estimates of genetic diversity in the field, further confirmation that the population numbers are critically low (Saarinen et al. 2009). Demographic stochasticity (including the timing and ability of adults to find mates and the influence of the Allee effect) and genetic stochasticity are highlighted as important factors in determining the future of this butterfly.

Building off of the success of amplifying microsatellite DNA from museum specimens (Meglécz et al. 1998), Saarinen and Daniels (2012) performed an evaluation of genetic diversity in a historical population of Miami blue butterflies from a large museum collection. Using modified ancient DNA techniques, we showed that genetic diversity was historically high in a population of Miami blue butterflies in 1940, 1960, and 1980, and that genetic diversity decreased over time. Historical populations were consistently in Hardy-Weinberg equilibrium, while current populations were not. Historical populations had a higher number of effective alleles (at the single microsatellite locus that could be amplified from older specimens) as well as a broader distribution of allelic frequency (more mid-frequency alleles) as compared with the extant populations (Saarinen and Daniels 2012). By evaluating a population over several decades, this study provides a picture of the way genetic diversity is distributed and lost over time in a large population.

It also serves as a benchmark for a level of genetic diversity known to support a large population of Miami blue butterflies. This is useful for management plans in evaluating current levels of genetic diversity and determining how far from historical measures the current population may be. These data will help in setting reasonable, meaningful management goals as recovery efforts continue.

Summary and Key Points

One lesson I have learned from conservation genetics studies is the importance of adequate sampling. By this, I mean that robust, statistically-relevant data are produced when large samples are assessed. This may appear as a bit of a catch-22, because in order to numerically sample a large number of individuals, we would need to have a large number of individuals in the first place. When dealing with threatened and endangered species they are, by definition, not abundant and therefore we cannot gather many samples. Therefore, I advocate for several solutions. First; I advocate for multiple genetic markers to be used to quantify the genetic diversity of the species' genome. This has statistical relevance (Ryman et al. 2006) as well as the utility of providing information on neutral and (potentially, if proper markers are chosen) fitness-related portions of the genome. One way to accomplish this would be to use multiple microsatellite markers (or combinations of different data types, see also Forister et al. 2008).

Second, I encourage that the entire species range be sampled for genetic work. This will provide the best picture of diversity that will be most useful for conservation planning. This may provide information on novel sources of diversity (from isolated populations) that could be useful for captive propagation work. Additionally, this provides a comparison of diversity from the center of the habitat vs. the range margins. Again, this may be useful in comparisons of habitat diversity and genetic diversity, which may be considered in management or reintroduction programs. More samples from more areas will allow us to identify genetic diversity and hopefully will allow us to take measures to preserve it.

Building from the spatial sampling, I also support temporal sampling when possible. I advocate for the assessment of museum specimens but only when historical series of the species are available (e.g. 20–30 samples from the same location and collection time). I do not advocate for museum work to be conducted on small sample sizes, as this will not provide an accurate assessment of historical genetic diversity. It will also use valuable research funds and time. However, when historic population genetic signatures can be recreated, they may provide a useful benchmark to compare current populations. They may also provide useful conservation genetics targets (i.e. maintenance of a percentage of historical diversity).

Successful projects should also strive for as many polymorphic, informative markers as possible, as these will increase the statistical signal and provide clearer results. Several successful projects (i.e. the Karner blue) have used multiple marker types to evaluate the species across its range, and these markers have provided answers to questions of conservation concern.

The issue of pathogen infection and spread should not be discounted or overlooked, especially when considering programs that propose captive breeding, translocation, or assisted migration. Again, several case studies have shown that butterflies are readily infected by the bacteria *Wolbachia* and this may have negative consequences for populations.

Lastly, we cannot discount a species genetic potential and ability to adapt to climate change. Although beyond the scope of this chapter, we do need to be mindful of populations remaining genetic diversity and how it will serve future generations as they face changing precipitation and temperature patterns as already seen in the study by Zakharov and Hellmann (2008). As we continue to develop new genetic and genomic methods (and prices for these services continue to fall), we may soon be able to conduct more analysis of species functional genetic variation, especially as it relates to their abilities to manage pathogen load and a changing climate (Kohn et al. 2006).

Overview of How to Begin a Conservation Genetics Project

While it is recommend that you start at step 1 with the determination of questions, the other actions should be considered in concert with each other. I hope that the end result will be a series of action items that will be executable based on a cohesive plan, complete with contingencies.

1. Assemble stakeholders.
2. Determine the questions that need to be addressed (brainstorm).
3. Rank the level of importance of the outcomes to the questions.
    (a) Are these results going to be included in management plans? Concentrate on questions where different outcomes would result in different management plan actions.
    (b) Are some questions more time-sensitive than others? Sometimes management plan deadlines and public interest may promote certain management strategies. For example, if translocation of individuals is proposed, pathogen screening should be prioritized.
4. Consider the feasibility of the questions from the organisms' perspective.
    (a) Do you have access to sample enough individuals? Statistical power is a very important consideration in studies of endangered species.
    (b) Will your sampling negatively impact sensitive habitat?
    (c) Will your sampling potentially endanger the population?
    (d) Consider how this may be mitigated.

5. Consider the feasibility of the questions from an agency perspective.
   (a) Does your agency have the time, money, expertise to perform these tasks? Are there partnerships with academic institutions?
   (b) Consider how this may be mitigated.
6. What molecular methods are available? What caveats do I need to consider?

This is an iterative process. While priorities are likely to change in multiple scenarios, having all questions accumulated at the outset will help determine the best course of action and promote the best genetic plan and will save time and money in the long run.

Acknowledgements I would like to thank the members of several working groups for their cooperation and feedback in the conservation genetics of endangered butterflies, especially the Karner Blue Butterfly Working Group, the Miami Blue Butterfly Working Group, the Poweshiek Skipperling and Dakota Skipper Working Group, the Mitchell's Satyr Working Group, and the Minnesota Zoo (Erik Runquist). Additional thanks for assistance with genetic research related to the Karner blue butterfly to Maria Albright (Allegan State Game Area, Michigan), Peter Tolson, Mitchell Magditch, and Candee Ellsworth (Toledo Zoo), Jason Dzurisin and Jessica Hellmann (University of Notre Dame), and Tim King (USGS, West Virginia). We thank David Cuthrell (Michigan Natural Features Inventory, Michigan State University) for field work collaboration, feedback on tissue collection methods, and photography. We thank Charlotte Germain-Aubrey for providing comments on the manuscript. Portions of the research presented in this chapter were supported by grants to E. Saarinen from the US Fish and Wildlife Service and Canon National Parks Science Scholars Program. Lastly, we thank all of the dedicated Lepidopterists who have carefully collected field data and tissue samples for the preservation of these species.

References

Amos W, Balmford A (2001) When does conservation genetics matter? Heredity 87:257–265

Avise JC (1994) Molecular markers, natural history, and evolution. Chapman & Hall, New York

Baldo L, Hotopp JCD, Jolley KA et al (2006) Multilocus sequence typing system for the endosymbiont Wolbachia pipientis. Appl Environ Microbiol 72:7098–7110

Brakefield P (1998) The evolution–development interface and advances with the eyespot patterns of Bicyclus butterflies. Heredity 80:265–272

Calhoun JV, Slotten JR, Salvato MH (2002) The rise and fall of tropical blues in Florida: Cyclargus ammon and Cyclargus thomasi bethunebakeri (Lepidoptera: Lycaenidae). Holarct Lepid 7(1):13–20

Caplins SA, Gilbert KJ, Ciotir C et al (2014) Landscape structure and the genetic effects of a population collapse. Proc R Soc B. doi:10.1098/rspb.2014.1798

Cassel A, Tammaru T (2003) Allozyme variability in central, peripheral and isolated populations of the scarce heath (Coenonympha hero: Lepidoptera, Nymphalidae): implications for conservation. Conserv Genet 4:83–93

Costanza R, d'Arge R, de Groot R et al (1997) The value of the world's ecosystem services and natural capital. Nature 387:253–260

Crawford LA, Koscinski D, Watt KM et al (2013) Mating success and oviposition of a butterfly are not affected by non-lethal tissue sampling. J Insect Conserv 17(5):859–864

Daniels JC, Saarinen EV, Emmel TC (2008) Molecular diversity of the state-endangered Miami blue butterfly. Final Report to the Florida Fish and Wildlife Conservation Commission. Tallahassee, Florida

DeSalle R, Amato G (2004) The expansion of conservation genetics. Nat Rev Genet 5:702–712

DOT 173.4. Title 49 CFR, subtitle B, chapter 1, subchapter C: Hazardous materials regulations. US Department of Transportation (DOT). http://www.access.gpo.gov/nara/cfr/waisidx_99/49cfrv2_99.html

Duron O, Buchon D, Boutin S et al (2008) The diversity of reproductive parasites among arthropods: Wolbachia do not walk alone. BMC Biol. doi:10.1186/1741-7007-6-27

Dyson EA, Kamath MK, Hurst GD (2002) Wolbachia infection associated with all-female broods in Hypolimnas bolina (Lepidoptera: Nymphalidae): evidence for horizontal transmission of a butterfly male killer. Heredity 88:166–171

Excoffier L, Lischer HEL (2010) Arlequin suite ver 3.5: a new series of programs to perform population genetics analyses under Linux and Windows. Mol Ecol Res 10:564–567

Forister ML, Nice CC, Fordyce JA et al (2008) Considering evolutionary processes in the use of single-locus genetic data for conservation, with examples from the Lepidoptera. J Insect Conserv 12:37–51

Forister ML, Gompert Z, Fordyce JA et al (2011) After 60 years, an answer to the question: what is the Karner blue butterfly? Biol Lett 7(3):399–402

Frankham R (1995) Conservation genetics. Annu Rev Genet 29:305–327

Frankham R, Ballou JD, Briscoe DA (2003) Introduction to conservation genetics. Cambridge University Press, Cambridge

Goldstein PZ, DeSalle R, Amato G et al (2000) Conservation genetics at the species boundary. Conserv Biol 14:120–131

Gompert Z, Nice CC, Fordyce JA et al (2006) Identifying units for conservation using molecular taxonomy: the cautionary tale of the Karner blue butterfly. Mol Ecol 15:1759–1768

Gompert Z, Forister ML, Fordyce JA et al (2010) Bayesian analysis of molecular variance in pyrosequences quantifies population genetics structure across the genome of Lycaeides butterflies. Mol Ecol 19:2455–2473

Hamm, CA (2012) Natural history and conservation genetics of the federally endangered Mitchell's Satyr butterfly, Neonympha mitchellii mitchelli. Dissertation, Michigan State University

Hamm CA, Aggarwal D, Landis DA (2010) Evaluating the impact of non-lethal DNA sampling on two butterflies, Vanessa cardui and Satyrodes eurydice. J Insect Conserv 14:11–18

Hamm CA, Rademacher V, Landis DA et al (2014a) Conservation genetics and the implication for recovery of the endangered Mitchell's satyr butterfly, Neonympha mitchelli mitchelli. J Hered 105(1):19–27

Hamm CA, Handley CA, Pike A et al (2014b) Wolbachia infection and Lepidoptera of conservation concern. J Insect Sci 14(6):1–8

Hiroki M, Tagami Y, Miura K et al (2004) Multiple infection with Wolbachia inducing different reproductive manipulations in the butterfly Eurema hecabe. Proc R Soc Lond B 271:1751–1755

Jennings TN, Knaus B, Mullins TD et al (2011) Multiplexed microsatellite recovery using massively parallel sequencing. Mol Ecol Res 11(6):1060–1067

Jiggins FM, Hurst GDD, Majerus MEN (1998) Sex ratio distortion in Acraea encedon (Lepidoptera: Nymphalidae) is caused by a male-killing bacterium. Heredity 81:87–91

Jiggins FM, Hurst GDD, Jiggins CD et al (2000) The butterfly Danaus chrysippus is infected by a male-killing Spiroplasma bacterium. Parasitology 120:439–446

Kearse M, Moir R, Wilson A et al (2012) Geneious Basic: an integrated and extendable desktop software platform for the organization and analysis of sequence data. Bioinformatics 28(12):1647–1649

Kohn MH, Murphy WJ, Ostrander EA et al (2006) Genomics and conservation genetics. Trends Ecol Evol 21(11):629–637

Koscinski D, Crawford LA, Keller HA et al (2011) Effects of different methods of non-lethal tissue sampling on butterflies. Ecol Entomol 36:301–308

Laikre L, Allendorf FW, Aroner LC et al (2009) Neglect of genetic diversity in implementation of the convention on biological diversity. Conserv Biol 24(1):86–88

Meglécz E, Pecsenye K, Varga A et al (1998) Comparison of differentiation pattern at allozyme and microsatellite loci in Parnassius mnemosyne (Lepidoptera) populations. Hereditas 128:95–103

Meglécz E, Petenian F, Danchin E et al (2004) High similarity between flanking regions of different microsatellites detected within each of two species of Lepidoptera: Parnassius apollo and Euphydryas aurinia. Mol Ecol 13:1693–1700

Morin PA, Martien KK, Archer FI et al (2010) Applied conservation genetics and the need for quality control and reporting of genetic data used in fisheries and wildlife management. J Hered 101(1):1–10

Moscardi F (1999) Assessment of the application of baculoviruses for control of Lepidoptera. Ann Rev Entomol 44:257–289

Nabokov V (1949) The Nearctic members of Lycaeides Hubner (Lycaenidae, Lepidoptera). Bull Mus Comp Zool 101:479–541

Nabokov V (1975) From letter to Robert Dirig. In: Boyd B, Pyle RM (eds) Nabokov's butterflies. Beacon, Boston

Nice CC, Anthony N, Gelembiuk G, Raterman D, Ffrench-Constant R (2005) The history and geography of diversification within the genus Lycaeides in North America. Mol Ecol 14:1741–1754

Nice CC, Gompert Z, Forister ML et al (2009) An unforeseen foe in arthropod conservation efforts: the case of Wolbachia infection in the Karner blue butterfly. Biol Conserv 142:3137–3146

Packer L, Taylor JS, Savignano DA et al (1998) Population biology of an endangered butterfly, Lycaeides melissa samuelis (Lepidoptera: Lycaenidae): genetic variation, gene flow, and taxonomic status. Can J Zool Rev Can Zool 76:320–329

Peakall R, Smouse PE (2012) GenAlEx 6.5: genetic analysis in Excel. Population genetic software for teaching and research-an update. Bioinformatics 28:2537–2539

Pritchard JK, Stephens M, Donnelly P (2000) Inference of population structure using multilocus genotype data. Mol Biol Evol 15:1298–1311

Ryman N, Palm S, André C et al (2006) Power for detecting genetic divergence: differences between statistical methods and marker loci. Mol Ecol 15:2031–2045

Saarinen, EV (2009) Population genetics of the endangered Miami blue butterfly Cyclargus thomasi bethunebakeri: implications for conservation. Dissertation, University of Florida

Saarinen EV, Austin JA (2010) When technology meets conservation: increased microsatellite marker production using 454 genome sequencing on the endangered Okaloosa darter (Etheostoma okaloosae). J Hered 101:784–788

Saarinen EV, Daniels JC (2012) Using museum specimens to assess historical distribution and genetic diversity in an endangered butterfly. Anim Biol 62(3):337–350

Saarinen EV, Daniels JC, Maruniak JE (2009) Development and characterization of polymorphic microsatellite loci in the endangered Miami blue butterfly Cyclargus thomasi bethunebakeri (Lepidoptera: Lycaenidae). Mol Ecol Res 9:242–244

Saarinen EV, Austin JD, Daniels JC (2010) Genetic estimates of contemporary effective population size in an endangered butterfly indicate a possible role for genetic compensation. Evol Appl 3:28–39

Saarinen EV, Daniels J, Maruniak J (2014) Local extinction despite high levels of gene flow and genetic diversity in the federally-endangered Miami blue butterfly. Conserv Genet 15:811–821

Scriber JM, Ording GJ, Mercader RJ (2008) Hybrid introgression and parapatric speciation in a hybrid zone. In: Tilmon KJ (ed) Specialization, speciation, and radiation: the evolutionary biology of herbivorous insects. University of California Press, Berkeley

Turlure C, Vandewoestijne S, Baguette M (2014) Conservation genetics of a threatened butterfly: comparison of allozymes, RAPDs, and microsatellites. BMC Genet. doi:10.1186/s12863-014-0114-7

USPS Publication 52. Hazardous, restricted, and perishable mail. Part 3: Hazardous materials, pp 15–60, July 1999. http://www.usps.com/cpim/ftp/pubs/pub52.pdf

Van Oosterhout C, Hutchinson WF, Wills DPM et al (2004) MICRO-CHECKER: software for identifying and correcting genotyping errors in microsatellite data. Mol Ecol Res 4(3):535–538

Vernesi C, Bruford MW, Bertorelle G et al (2008) Where's the conservation in conservation genetics? Conserv Biol 22(3):802–804

Werren JH (1997) Biology of Wolbachia. Ann Rev Entomol 42:587–609

Wheat CW, Haag CR, Marden JH et al (2010) Nucleotide polymorphism at a gene (Pgi) under balancing selection in a butterfly metapopulation. Mol Biol Evol 27(2):267–281

Wiemers M, Fiedler K (2007) Does the DNA barcoding gap exist? – a case study in blue butterflies (Lepidoptera: Lycaenidae). Front Zool. doi:10.1186/1742-9994-4-8

Zakharov EV, Hellmann JJ (2008) Genetic differentiation across a latitudinal gradient in two co-occurring butterfly species: revealing population differences in a context of climate change. Mol Ecol 17(1):189–208

Zhan S, Zhang NK et al (2014) The genetics of monarch butterfly migration and warning colouration. Nature 514:317–321

Zhang DX (2004) Lepidopteran microsatellite DNA: redundant but promising. Trends Ecol Evol 19(10):507–509

Zhou WG, Rousset F, O'Neill S (1998) Phylogeny and PCR-based classification of Wolbachia strains using wsp gene sequences. Proc R Soc Lond B 265:509–515, Los Angeles River Watershed. University of Southern California Center for Sustainable Cities, Los Angeles

Managing Land for Butterflies

Doug Taron

Abstract Because butterflies are important contributors to biodiversity of natural areas, their habitat needs should be considered when making decisions regarding land management. In order to do this effectively it's important to know what factors are essential components of butterfly habitat. In addition to host plants and nectar resources, a wide variety of parameters, including vegetative structure, presence of mutualist organisms, and edaphic as well as other factors all contribute to the overall quality of butterfly habitat. Population structure is also an important consideration, particularly in cases where metapopulation dynamics are present. Land managers can simplify their task considerably by limiting their considerations to those species that require the habitats that are under active management while recognizing that species that are adapted to the broader landscape are much less likely to be affected by management activities.

Management of natural areas is typically important for maintaining butterfly populations and for controlling invasive species, many of which have deleterious effects on butterfly populations. Decisions regarding the application of management techniques, particularly fire, can be challenging, in part because of the range of opinions held by researchers regarding both optimal implementation of these techniques and their effects on butterflies. In general, land managers should apply management techniques, including fire, as conservatively as possible while being consistent with management objectives.

Introduction

In an ideal world, managing land for butterflies would incorporate detailed knowledge of a species' habitat requirements with a suite of management techniques that can be applied to enhance those parameters without otherwise damaging the butterfly population. Unfortunately, neither this sort of detailed knowledge nor ideal management techniques are reliably available in the real world. In the face

---

D. Taron (✉)
The Chicago Academy of Sciences, Peggy Notebaert Nature Museum, Chicago, IL, USA
e-mail: dtaron@naturemuseum.org

© Springer Science+Business Media B.V. 2015
J.C. Daniels (ed.), *Butterfly Conservation in North America*,
DOI 10.1007/978-94-017-9852-5_6

of incomplete knowledge and imperfect management techniques, there are none the less many avenues open to landowners for more effectively managing sites for butterfly conservation.

Habitat and Fragmentation

Defining and Modeling Habitat

Habitat for many species remains, at best, very generally described. In practice, land managers frequently define habitat in terms of affiliation with a particular plant community such as salt marsh, sparsely vegetated sand dunes, or tallgrass prairie. Habitat modeling approaches attempt to apply additional rigor to the definition of habitat for a particular species (Dennis et al. 2006). Such approaches typically require the collection of a considerable body of data that is used in constructing a model of a species' habitat. Variables addressed by the model typically incorporate far more than abundance and distribution of host plants and nectar resources:

> The more resources are taken into account in defining habitats in butterfly biology (e.g., thermoregulatory substrates, mate location sites, nectar, roosts, hibernation sites), as well as host plants, the greater the number of resources that will be found within the matrix. (Dennis et al. 2006)

Detailed habitat models are species- and, frequently, site-specific. Consequently, they are available for only a very small number of especially well-studied species (Schultz and Crone 1998, 2005; Schultz and Hammond 2003; Schtickzelle et al. 2005; Grundel and Pavlovic 2007). While potentially of significant value in advancing understanding of the ecology of imperiled species, habitat models do not always provide the kinds of recommendations that are of use to land managers. For example, many habitat models are effective in predicting the presence or absence of a species of interest without offering the kinds of specific guidance that can be applied to facilitate management for that species.

While habitat models can provide excellent insights into the needs of imperiled butterflies, such models are likely to remain beyond the scope of what can be undertaken by many land managers. This is especially true in situations that involved declining, though not yet listed species. Although it can be very desirable to take steps to prevent continued decline to the point that listing is warranted, such efforts can be hampered by limited resources. Additionally, land management typically involves attempts to protect a wide range of taxa, and often emphasizes sites' floristic qualities.

Fortunately, there is some evidence that designing and managing pre-serves predominantly based on their vegetative composition can be reasonably effective in conserving insects, including butterflies (Panzer and Schwartz 1998).

Matrix and Island Species

The challenges of managing land in a way that increases (or at least maintains) populations of numerous butterfly species, many with different ecological needs, can appear daunting. If that weren't enough, land is typically managed with at least an eye towards the ecological requirements of other taxa as well. One way to narrow this problem to a more tractable level is to be mindful of the range of conservation needs of different species.

Numerous authors have used a variety of terminology to distinguish matrix and island species (New et al. 1995; Panzer et al. 1995). The latter are, to some degree, restricted to "islands" of native or semi-native vegetative communities. Such communities are frequently the focus of land management activities. In contrast, matrix species can inhabit the more altered vegetation that surrounds these islands, though these species are frequently also found on the islands. Especially in more highly fragmented landscapes, management activities on islands of remnant vegetation may have little or no effect on populations of butterflies that inhabit the matrix. In cases where management of a remnant of native vegetation diminishes a species that also resides in the matrix, the matrix can be expected to provide individuals to repopulate the island. The converse is generally not true, though repopulation from other nearby islands sometimes remains a possibility, as will be discussed further below.

Island species have been observed to represent a minority of the taxa present in a given region. For example, one study in the Chicago region (Panzer et al. 1995) found that just 25 % of insect species (though 40 % of butterfly species) showed a strong fidelity to the region's prairie and savanna remnants. These authors point out that the Chicago region likely represents an extreme in landscape fragmentation, and suggested that in less severely fragmented prairie and savanna remnants further to the west, a smaller proportion of species may show a comparably high level of fidelity.

By focusing the management decision process around island species, land managers can target resources to those species that are most in need of conservation attention. For example, between 30 and 50 species might breed on a 50-acre high quality prairie remnant in Illinois, however very few such sites support more than a dozen or so island species (Illinois Butterfly Network, pers. comm.). It is important to note that certain species show different island and matrix affinities in different parts of their ranges. For example, the Aphrodite fritillary (*Speyeria aphrodite*) is predominantly restricted to larger remnant prairies in Illinois (Panzer et al. 1995; Illinois Butterfly Network, pers. comm.), whereas Wisconsin populations are found in a much wider variety of grassland and savanna habitats (Swengel 1997). Thus, the notion of island and matrix species, while useful in guiding management decisions, should employ region-specific information regarding the fidelity of various taxa to remnant vegetation.

Metapopulations

Butterflies were among the taxa used to first elucidate the concept of metapopulations (Ehrlich and Murphy 1987; Harrison et al. 1988; Hanski et al. 1994), a concept that remains important in their conservation. Metapopulations can be thought of as populations of populations. Each subpopulation, or deme, within a metapopulation is at least semi-isolated from other subpopulations such that gene flow within a given deme is typically greater than gene flow between demes. Metapopulations are frequently highly dynamic systems in which given demes may undergo periodic local extinction followed by repopulation via migration from other demes within the metapopulation.

The metapopulation model is important for both the design and management of preserves that are intended to conserve butterfly populations (Schtickzelle et al. 2005). Because of their dynamic process of extinctions and recolonizations, metapopulations can be substantially more persistent than individual demes. In cases where preservation or management activities do not take into account the full extent of the metapopulation, these efforts are less likely to succeed in the long run than activities that include the entire metapopulation.

Thus, in order to maximize the likelihood of long-term persistence of butterfly metapopulations, it is incumbent on land managers to develop knowledge of the full extent of a metapopulation and at least the rough locations of the individual demes therein. A significant challenge to land managers arises out of the species-specific nature of metapopulation dynamics (Baguette et al. 2000). This challenge is most acute in situations where multiple species of interest are present on the same parcel of land. Identifying and heeding the population structure of just one of these species will not necessarily ensure adequate protection of the others.

Components of Habitat and Their Management

Resources

Dennis et al. (2006) divide resources that butterfly species require from their habitats into consumables, such as host plants and nectar sources, and utilities. Each of these resources needs present both challenges and opportunities for land managers.

Host Plants: The specificity of the relationship between butterfly larvae and their host plants drives a lot of the thinking regarding preserve design and management. Although presence of host plants is a necessary condition for supporting a breeding population of butterflies, it is not always sufficient. Both the abundance and distribution of host plants factor in to the suitability of a given patch.

In the case of Oregon silverspot (*Speyeria zerene hippolyta*) butterflies, individual Viola host plants are insufficient to support the full development of the larvae to the pupal stage. Moreover, foraging larvae were shown to be unable to distinguish their Viola host plants from surrounding vegetation, and thus depend on random wandering rather than any sort of directed or guided search (Bierzychudek et al. 2009).

This finding suggests that a certain minimum density of host plants is required to sustain a viable population. Thus, in at least some cases it may be more desirable to establish host plants as discontinuous dense patches rather than continuously but at lower density across the full range of a site. On a site of sufficiently large size, this approach has the added advantage of the possibility of the establishment of a metapopulation dynamic and the accrual of the enhanced persistence of populations that display this dynamic.

Nectar Resources: In general, there is much less species-specificity for nectar resources than for host plants. Many butterflies tend to be opportunistic, utilizing an array of flowering plant species, provided they can access the available nectar. Although there have been suggestions regarding floral color preferences in certain butterfly species (Ômura and Honda 2005), many plants are good nectar resources for a relatively wide range of butterfly species, for example plants in the genera *Asclepias, Helianthus,* or *Eupatorium.*

The need for synchrony between nectar resources and the adult flight period should not be overlooked, particularly in the case of multivoltine species such as the Karner Blue (*Lycaeides melissa samuelis*) (Grundel and Pavlovic 2007) or species with prolonged adult flight periods, such as fritillaries in the genus *Speyeria*, which may begin flying in June but delay oviposition until late in the summer (Powell et al. 2007). As a general rule of thumb, land managers wishing to maximize butterfly diversity on their sites should manage for abundant nectar sources throughout the growing season.

Other Mutualistic Organisms: An often overlooked resource required by certain species of butterflies are mutualistic organisms, in particular the various species of ants that engage in symbiotic relationships with butterflies in the family Lycaenidae. These butterflies include imperiled species such as the Karner blue (Savignano 1994; Grundel et al. 1998) and Miami blue (*Cyclargus thomasi bethunebakeri*) (Saarinen and Daniels 2006; Trager and Daniels 2009). Steiner et al. (2003) advised managers to identify host ant species "in order to avoid failure of management programs or even damage to populations on the edge of extinction."

Abiotic Factors

Factors other than resources can limit the ability of butterflies to occupy certain sites.

Vegetation Structure: Some species are absent from certain sites that appear to contain adequate resources. Regal fritillaries for example, can be reared on a variety of violet species in the laboratory, however in the field they are not found in many situations, such as open woodlands, with abundant host plants and nectar resources (Ries and Debinski 2001; Ries et al. 2001). Invasion of remnant grasslands in western Oregon by tall oat grass has led to changes in butterfly fitness and various fitness-related behaviors in endangered Fender's blues (Severns 2008). Microclimate can also be affected be vegetative structure, as illustrated by the preference of green hairstreaks for the transition zone between forest and heathland, despite the abundance of both host plant and nectar resources in the more open adjacent heathland (Vanreusel and Van Dyck 2007). Silver-studded blues in coastal Britain favored shrubby patches of their herbaceous host plants over more open patches. The shrubby vegetative structure allowed them to change their roosting, perching, and basking behavior as weather conditions changed (Dennis and Sparks 2006). Shrub patches in otherwise open prairies are used by regal fritillaries during summer diapause (Taron, pers. obs.).

Edaphic Factors: Many prairie skippers are substantially less common than their host plants, which sometimes include relatively abundant prairie grasses. These skippers are absent from many sites that have abundant host plant resources. Abiotic factors are hypothesized to be responsible however in most cases the details have not been elucidated. An exception is the imperiled Dakota skipper (*Hesperia dacotae*). The species is restricted to specific types of glacial deposits of sand and gravel, where its distribution is further circumscribed by edaphic factors. Altered vertical distribution of water in the soil appears to leave larvae vulnerable to desiccation in late summer (Royer et al. 2008).

Reserve Size, Isolation, and Shape: Regardless of whether a population of butterflies is a deme in a metapopulation or a stand-alone population, there is a lower limit to the area that will support the deme or population. Various metapopulation studies have documented a large range of patch sizes, ranging from 0.1 to 100 ha (Dennis et al. 2006). In cases of isolated populations, the lower limit is likely to be considerably larger. In the Chicago area, Panzer et al. (1995) surveyed species richness among butterfly species that rely on upland prairies and found a strong area effect on sites that ranged from 2 to 60 ha.

Both patch isolation and patch shape contribute to the risk of elimination of species from patches of habitat. Risk of extirpation generally increases with increasing isolation (Dover and Settele 2009). A study on uncultivated portions of organic farms in Denmark showed greater species richness on more compact patches compared with more linear patches (Clausen et al. 2001). In summary, larger, more connected, and more compact reserves show greater resilience to loss of butterfly species than do reserves that are smaller, more isolated, and more linear.

Surrounding Matrix: It's important to consider the role of the matrix on habitat patch occupancy (Sweaney et al. 2014). Surrounding matrix influences butterfly species composition of habitat islands (Bergman et al. 2004; Öckinger et al. 2012) and can facilitate or hinder the dispersal of butterflies between patches (Ries and Debinski 2001; Davis et al. 2007). The matrix is not necessarily empty with regard to species of concern, nor is it devoid of resources for butterfly species characteristic of remnants. Rather remnant-associated butterflies are present there at far lower densities and frequencies than they are in the habitat patches. Research suggests that by enhancing matrix areas within a reserve, such as by promoting nectar resources within degraded areas, can enhance the mobility of butterflies between patches of habitat (Ries and Debinski 2001).

Integrating Management Practices and Conservation: What's a Land Manager to Do?

Preservation Isn't Enough

The idea that setting aside a reserve without managing it will not preserve biodiversity has been long established relative to plant communities. In particular, grassland and savanna communities are vulnerable to displacement by woody vegetation through natural successional processes. Because of the damage that these processes can do to plant communities, they can also damage populations of butterflies that depend on these communities.

Elimination of Habitat through Natural Succession: Changes to sub-climax plant communities, such as prairies, through natural succession, is well documented (Bragg and Hulbert 1976; McClain and Anderson 1990; Kline 1997), as are associated changes to butterflies (New 1991). Displacement by encroaching woody vegetation has been documented for diverse species, including fen specialist butterfly species in Michigan (Summerville and Clampitt 1999), regal fritillaries over much of the eastern United States (Ferster and Vulinec 2010), and grassland butterflies from the Willamette Valley in Oregon (Schultz et al. 2011). The Oregon species were also affected by invasion of tall non-native grasses.

Invasive Species: In addition to their negative effects on plant communities, invasive species have various deleterious effects on butterfly populations. These effects were recently reviewed by Wagner and Van Driesche (2010). In some cases these effects are due to the changes in floristic composition and structure that result from invasion by alien species.

Other invasive plant species can act as evolutionary traps (Schlaepfer et al. 2005) by, for example, providing stimuli for oviposition but being poorly able to support larval development. A significant invasive weed of woodlands, garlic mustard, has been shown to represent such an ecological trap for the West Virginia white (Bowden 1971; Keeler and Chew 2008), which has exhibited population declines as a result. A similar pattern of poor larval survival following oviposition has been observed between monarch larvae and the invasive plant black swallow-wort (DiTommaso and Losey 2003).

Certain control measures against invasive species also have a history of deleterious effects on native lepidopteran populations. Here the issue is less with invasive plants than with invasive insects, particularly gypsy moths (Wagner and Van Driesche 2010). Aerial spraying of the Lepidoptera-specific insecticide Btk has caused localized extirpations of populations of several species including the Appalachian grizzled skipper (*Pyrgus wyandot*), (Schweitzer 2004). Severns (2002) reported localized extinctions of three species of butterflies following BTk application to a park in western Oregon. The tachinid fly *Compsilura concinnata*, introduced as a gypsy moth control agent, has been implicated in the decline of saturniid moths in New England (Boettner et al. 2000).

Control Measures

One of the significant challenges to the evaluation of measures recommended to maintain subclimax ecosystems is the large range of often contradictory findings relative to the effects of management techniques on butterflies. Although many of the same processes that degrade plant communities are also deleterious to the butterflies that inhabit these communities, measures taken to improve habitat may have different effects on native plants and butterflies.

Several studies have suggested that management that benefits plant communities can also benefit butterfly populations. For example, occupancy of habitat patches by the large heath (*Coenonympha tullia*) was affected by habitat quality as much as it was by patch size or isolation (Dennis and Eales 1997). Other studies have shown management to be both compatible with (Dollar et al. 2013) and deleterious to (Swengel et al. 2011) butterfly populations.

A variety of techniques, including burning, mowing, or grazing has been used to prevent the invasion of grassland communities by woody vegetation. The effects of these practices have mostly been studied as compared with those of burn management (Moranz et al. 2012). Vogel et al. (2007) evaluated the effects of various patterns of management on butterfly species richness in grassland habitats, finding that species responded differently to different management practices, stating that "... different restoration practices tend to favor some species and not others even within the same habitat guild." Recently Moran (2014) suggested that grazing with bison is correlated with increased arthropod diversity.

Fire

For many land managers, questions regarding the use of prescribed fire and its effects on butterflies are pressing. Whereas the importance of fire as a management tool is widely recognized (Gleason 1913; Anderson 1990; Pillsbury et al. 2011; Arthur et al. 2012), many land managers are also aware of the potential for this practice to harm populations of insect species.

Numerous studies address the effects of fire on insects (Reed 1997; Panzer 2002; Dollar et al. 2013; Swengel 2001). This literature makes a variety of recommendations, and there seems little consensus regarding optimal approaches to burning. Land managers can understandably feel confused regarding the development of burn plans that avoid eliminating important insect species. Understanding how fire affects butterflies and other insects may provide guidance in tailoring a burn regimen to a particular situation.

How Burns Affect Insects

There are several phases in the ecological response to a burn, each having a distinct set of effects on insect populations (Reed 1997). The most immediate effects occur in the combustion phase, when the most direct mortality occurs. In the shock phase immediately following a burn, little or no vegetation is present to support phytophagous insects such as butterfly larvae. There is less shelter from which insects can seek refuge from extremes of weather or predators. Other effects, both negative and positive, persist long after the burn itself during the recovery phase. As the recovery continues, deleterious effects of the burn diminish and benefits increase, as remaining insects are able to benefit by the stimulus that burning can provide to plant growth (Reed 1997; Panzer and Schwartz 2000; Panzer 2002; Dollar et al. 2013).

Not all insects respond equally to fire. Numerous authors have suggested that there is a range of fire tolerance and sensitivity (Panzer and Schwartz 2000; Panzer 2002; Swengel 2001; Huntzinger 2003). Certain species, for example moths in the genus *Schinia* (Hardwick 1958), overwinter underground as immature stages, and are therefore afforded considerable protection from burns that occur during the dormant season. Species that winter in standing vegetation, leaf litter, or duff are more vulnerable to direct effects of burning. There is evidence that species that rely on remnant native habitat are often the species that are most sensitive to the effects of fire (Swengel 1998, 2001). Specific examples include butterflies in the genus *Speyeria* (Swengel 1997; Swengel and Swengel 2001) and skippers (Schlicht and Orwig 1999; Schultz et al. 2011).

The sensitivity of particular insects to fire represents a serious conundrum for land managers. Certain fire-sensitive butterfly species, such as the Karner and Fender's blues, inhabit habitats that show a propensity to vanish through successional processes in the absence of fire (Shuey 1997; Schultz and Crone 1998). Habitats of these species could have persisted in a less fragmented landscape due to periodic fires that would have interrupted successional processes. Sensitive butterflies that inhabited these landscapes may have originally persisted as a result of metapopulation dynamics, whereby recovery from displacement due to fire could arise from demes persisting on adjacent unburned areas, or on unburned refugia within a burned area. Some support for this hypothesis has been provided by observations that prairie insects are rapidly able to re-colonize a burn from adjacent unburned areas (Panzer 2002).

This dynamic has important implications for land managers today, as habitat fragmentation frequently results in populations that are both smaller and more isolated than they were historically. This situation is particularly acute in circumstances where fragmentation has been the most severe, such as grasslands of the upper Midwest and West Coast. Burn frequencies that may be tolerated in a situation of lower habitat fragmentation may result in the elimination of a species in a more fragmented environment.

To Burn or Not to Burn?

Faced with the risks attendant with burning, how does one decide on an appropriate burn regimen for a given site? A wide range of studies and opinions has been published. The spectrum ranges from not burning at all, though burning at decadal or multi-decadal intervals, to burning at intervals of a few years (Schultz and Crone 1998; Swengel 1998; Panzer and Schwartz 2000; Panzer 2002; Dollar et al. 2013). In spite of the considerable range of opinion, there does appear to be general agreement that some of the insect species on managed preserves can be expected to be sensitive to fire, and that excessive burning can be deleterious.

The setting aside of permanent unburned refugia within preserves (Swengel and Swengel 2007) has been recommended as a means to minimize insect species loss, as has the application of a variety of management techniques such haying and mowing instead of (Swengel 1998) or in addition to (Vogel et al. 2007, 2010) prescribed burning. It may also be useful to avoid returning to ignite unburned areas that have been skipped by fires, as they might provide additional refugia for fire-sensitive insects (Taron 1996).

Specific recommendations for burn regimens are difficult to make. The same approaches are unlikely to be effective at preserving butterflies in all circumstances. Land managers are advised to burn as conservatively as is possible, while remaining consistent with maintaining vegetative structure and composition. The smaller and more isolated a site, the greater the imperative for managers to be conservative in their use of fire, particularly if sensitive species are known to inhabit these sites.

Burning in a way that avoids harm to butterfly populations will continue to represent a challenge to land managers for the foreseeable future. The effects of both prescribed and wild fires on insects remains an important research need, particularly as demand grows for increasingly specific guidance regarding the coexistence of insect conservation and use of fire as a management tool.

> Managers seeking to maximize the number of butterfly species could achieve this goal by using multiple types of management wherever possible to provide the different microhabitat types utilized by different butterfly assemblages. (Vogel et al. 2007)

Summary and Key Points

Faced with the possibility of conflicting management needs of different taxa that share the same site, how can land managers approach the challenge of managing these sites in a way that maximizes biodiversity? The way forward will necessarily require a delicate balance. Manage too vigorously and species could be lost quickly due to the immediate effects of excessive burning or ill-timed mowing. Insufficient management may result in a longer retention of butterfly species, but in the longer run the inexorable changes that result from ecological succession raises the risk that both butterfly and plant communities could be eliminated. In general, managers should:

- Be aware of the sensitive species on your site. By knowing which important species reside on a site, as well as what parts of the site they are using, managers can use particular caution on these more sensitive areas.
- Manage conservatively, particularly relative to the use of fire. What is the minimum amount of burning that is consistent with your management objectives?
- Try to use a variety of management techniques. Consider varying the use of fire, grazing, mowing, and hand clearing such that a given site does not receive only one approach to management in perpetuity.
- Plan for refugia for species that are sensitive to fire. Consider leaving areas as permanently unburned refugia. In addition, leave temporary refugia by not returning to light unburned areas where a fire has skipped.
- Take into account the size and isolation of your site, and the ability for re-colonization from nearby or adjacent unburned areas. Sites that are larger and more connected will show greater resilience. Sites that are very small may be insufficient to support species of butterflies that are of conservation concern. Midsized (50 to several hundred acres) sites that are isolated may represent a particular challenge in terms of managing for butterflies.
- When engaging in a complete ecosystem reconstruction, such as when creating a prairie on a field that was recently farmland, aggressive management is less likely to be harmful to sensitive species early in the process. Once a relatively diverse native matrix of plant species takes hold, it is likely to become progressively more important to consider other taxa in addition to the flora.

References

Anderson RC (1990) The historic role of fire in North American grassland. In: Collins SL, Wallace LL (eds) Fire in North American tallgrass prairies. University of Oklahoma Press, Norman, pp 8–18

Arthur MA, Alexander HD, Dey DC, Schweitzer CJ, Loftis DL (2012) Refining the oak-fire hypothesis for management of oak-dominated forests of the eastern United States. J For 110(5):257–266

Baguette M, Petit S, Quéva F (2000) Population spatial structure and migration of three butterfly species within the same habitat network: consequences for conservation. J Appl Ecol 37(1):100–108

Bergman KO, Askling J, Ekberg O, Ignell H, Wahlman H, Milberg P (2004) Landscape effects on butterfly assemblages in an agricultural region. Ecography 27:619–628

Bierzychudek P, Warner KA, Mchugh A, Thomas L (2009) Testing the host-finding ability of a monophagous caterpillar in the field. Ecol Entomol 34(5):632–637

Boettner GH, Elkinton JS, Boettner CJ (2000) Effects of a biological control introduction on three nontarget native species of saturniid moths. Conserv Biol 14:1798–1806

Bowden SR (1971) American white butterflies (Pieridae) and English food-plants. J Lepid Soc 25:6–12
Bragg TB, Hulbert LC (1976) Woody plant invasion of unburned Kansas bluestem prairie. J Range Manage 29(1):19–24
Clausen HD, Holbeck H, Reddersen J (2001) Factors influencing abundance of butterflies and burnet moths in the uncultivated habitats of an organic farm in Denmark. Biol Conserv 98(2):167–178
Davis JD, Debinski DM, Danielson BJ (2007) Local and landscape effects on the butterfly community in fragmented Midwest USA prairie habitats. Landsc Ecol 22:1341–1354
Dennis RLH, Eales H (1997) Patch occupancy in Coenonympha tullia (Müller 1764) (Lepidoptera: Satyrinae): habitat quality matters as much as patch size and isolation. J Insect Conserv 1:167–176
Dennis R, Sparks TH (2006) When is a habitat not a habitat? Dramatic resource use changes under differing weather conditions for the butterfly *Plebejus argus*. Biol Conserv 129(3):291–301
Dennis RLH, Shreeve TG, Van Dyck H (2006) Habitats and resources: the need for a resource-based definition to conserve butterflies. Biodivers Conserv 15:1943–1966
DiTommaso A, Losey JE (2003) Oviposition preference and larval performance of monarch butterflies (Danaus plexippus) on two invasive swallow-wort species. Entomol Exp Appl 108:205–209
Dollar JG, Riffell SK, Burger LW (2013) Effects of managing semi-natural grassland buffers on butterflies. J Insect Conserv 17:577–590
Dover J, Settele J (2009) The influences of landscape structure on butterfly distribution and movement: a review. J Insect Conserv 13:3–27
Ehrlich PR, Murphy DD (1987) Conservation lessons from long-term studies of checkerspot butterflies. Conserv Biol 1:122–131
Ferster B, Vulinec K (2010) Population size and conservation of the last eastern remnants of the regal fritillary, *Speyeria idalia* (Drury) [Lepidoptera, Nymphalidae]; implications for temperate grassland restoration. J Insect Conserv 14(1):31–42
Gleason HA (1913) The relation of forest distribution and prairie fires in the middle west. Torreya 13:173–181
Grundel R, Pavlovic NB (2007) Resource availability, matrix quality, microclimate, and spatial pattern as predictors of patch use by the Karner blue butterfly. Biol Conserv 135(1):135–144
Grundel R, Pavlovic NB, Sulzman CL (1998) The effect of canopy cover and seasonal change on host plant quality for the endangered Karner blue butterfly (*Lycaeides melissa samuelis*). Oecologia 114(2):243–250
Hanski I, Kuussaari M, Nieminen M (1994) Metapopulation structure and migration in the butterfly *Melitaea cinxia*. Ecology 75(3):747–762
Hardwick DF (1958) Taxonomy, life history, and habits of the elliptoid-eyed species of *Schinia* (Lepidoptera: Noctuidae), with notes on the Heliothidinae. Mem Entomol Soc Can 90(S6):5–116
Harrison S, Murphy DD, Ehrlich PR (1988) Distribution of the bay checkerspot butterfly, *Euphydryas editha bayensis*: evidence for a metapopulation model. Am Nat 132:360–382
Huntzinger M (2003) Effects of fire management practices on butterfly diversity in the forested western United States. Biol Conserv 113(1):1–12
Keeler MS, Chew FS (2008) Escaping an evolutionary trap: preference and performance of a native insect on an exotic invasive host. Oecologia 156:559–568
Kline VM (1997) Orchards of oak and a sea of grass. In: Packard S, Mutel CF (eds) The tallgrass restoration handbook for prairies, savannas, and woodlands. Island Press, Washington, DC
McClain WE, Anderson EA (1990) Loss of hill prairie through woody plant invasion at Pere Marquette State Park, Jersey County, Illinois. Nat Area J 10(2):69–75
Moran MD (2014) Bison grazing increases arthropod abundance and diversity in a tallgrass prairie. Environ Entomol 43(5):1174–1184

Moranz RA, Debinski DM, McGranahan DA, Engle DM, Miller JR (2012) Untangling the effects of fire, grazing, and land-use legacies on grassland butterfly communities. Biodivers Conserv 21:2719–2746

New TR (1991) Butterfly conservation. Oxford University Press, Oxford

New TR, Pyle RM, Thomas JA, Thomas CD, Hammond PC (1995) Butterfly conservation management. Ann Rev Entomol 40(1):57–83

Öckinger E, Bergman KO, Franzén M, Kadlec T, Krauss J, Kuussaari M, Bommarco R (2012) The landscape matrix modifies the effect of habitat fragmentation in grassland butterflies. Landsc Ecol 27(1):121–131

Ômura H, Honda K (2005) Priority of color over scent during flower visitation by adult Vanessa indica butterflies. Oecologia 142(4):588–596

Panzer R (2002) Compatibility of prescribed burning with the conservation of insects in small, isolated prairie reserves. Conserv Biol 16(5):1296–1307

Panzer R, Schwartz MW (1998) Effectiveness of a vegetation-based approach to insect conservation. Conserv Biol 12:693–702

Panzer R, Schwartz M (2000) Effects of management burning on prairie insect species richness within a system of small, highly fragmented reserves. Biol Conserv 96(3):363–369

Panzer R, Stillwaugh D, Gnaedinger R, Derkovitz G (1995) Prevalence of remnant dependence among the prairie- and savanna-inhabiting insects of the Chicago region. Nat Area J 15(2):101–116

Pillsbury FC, Miller JR, Debinski DM, Engle DM (2011) Another tool in the toolbox? Using fire and grazing to promote bird diversity in highly fragmented landscapes. Ecosphere 2(3):art28

Powell AF, Busby WH, Kindscher K (2007) Status of the regal fritillary (*Speyeria idalia*) and effects of fire management on its abundance in northeastern Kansas, USA. J Insect Conserv 11(3):299–308

Reed CC (1997) Responses of prairie insects and other arthropods to prescription burns. Nat Area J 17(4):380–385

Ries L, Debinski DM (2001) Butterfly responses to habitat edges in the highly fragmented prairies of Central Iowa. J Anim Ecol 70(5):840–852

Ries L, Debinski DM, Wieland ML (2001) Conservation value of roadside prairie restoration to butterfly communities. Conserv Biol 15(2):401–411

Royer RA, McKenney RA, Newton WE (2008) A characterization of non-biotic environmental features of prairies hosting the Dakota skipper (*Hesperia dacotae*, Hesperiidae) across its remaining US range. J Lepid Soc 62(1):1

Saarinen EV, Daniels JC (2006) Miami blue butterfly larvae (Lepidoptera: Lycaenidae) and ants (Hymenoptera: Formicidae): new information on the symbionts of an endangered taxon. Fla Entomol 89:69–74

Savignano DA (1994) Benefits to Karner blue butterfly larvae from association with ants. In: Andow DA, Baker RJ, Lane CP (eds) Karner blue butterfly: a symbol of a vanishing landscape. Minnesota Agricultural Experiment Station, University of Minnesota, St Paul, pp 37–46

Schlaepfer MA, Sherman PW, Blossey B, Runge MC (2005) Introduced species as evolutionary traps. Ecol Lett 8(3):241–246

Schlicht DW, Orwig TT (1999) The last of the Iowa skippers. Am Butterflies 7(1):4–13

Schtickzelle N, Choutt J, Goffart P, Fichefet V, Baguette M (2005) Metapopulation dynamics and conservation of the marsh fritillary butterfly: population viability analysis and management options for a critically endangered species in Western Europe. Biol Conserv 126:569–581

Schultz CB, Crone EE (1998) Burning prairie to restore butterfly habitat: a modelling approach to management tradeoffs for the Fender's blue. Restor Ecol 6:244–252

Schultz CB, Crone EE (2005) Patch size and connectivity thresholds for butterfly habitat restoration. Conserv Biol 19(3):887–896

Schultz CB, Hammond PC (2003) Using population viability analysis to develop recovery criteria for endangered insects: case study of the Fender's blue butterfly. Conserv Biol 17(5):1372–1385

Schultz CB, Henry E, Carleton A, Hicks T, Thomas R, Potter A, Reader B (2011) Conservation of prairie-oak butterflies in Oregon, Washington, and British Columbia. Northwest Sci 85(2):361–388

Schweitzer D (2004) Gypsy moth (*Lymantria dispar*): impacts and options for biodiversity-oriented land managers. NatureServe Explorer, Arlington, 59 pp

Severns P (2002) Evidence for the negative effects of Bt (*Bacillus thuringiensis var. kurstaki*) on a non-target butterfly community in western Oregon, USA. J Lepid Soc 56(3):166–170

Severns PM (2008) Exotic grass invasion impacts fitness of an endangered prairie butterfly, Icaricia icarioides fenderi. J Insect Conserv 12(6):651–661

Shuey JA (1997) Dancing with fire: ecosystem dynamics, management, and the Karner blue (*Lycaeides melissa samuelis* Nabokov) (Lycaenidae). J Lepid Soc 51:263–268

Steiner FM, Sielezniew M, Schlick-Steiner BC, Höttinger A, Stankiewicz H, Górnicki A. (2003) Host specificity revisited: new data on Myrmica host ants of the lycaenid butterfly Maculinea rebeli. J Insect Conserv 7:1–6

Summerville KS, Clampitt CA (1999) Habitat characterization of five rare insects in Michigan (Lepidoptera: Hesperiidae, Riodinidae, Satyridae: Homoptera: Cercopidae). Great Lakes Entomol 32(3):225–238

Sweaney N, Lindenmayer DB, Driscoll DA (2014) Is the matrix important to butterflies in fragmented landscapes? J Insect Conserv 18(3):283–294

Swengel AB (1997) Habitat associations of sympatric violet-feeding fritillaries (*Euptoieta*; *Speyeria*; *Boloria*) (Lepidoptera: Nymphalidae) in tallgrass prairie. Great Lakes Entomol 30:1–18

Swengel AB (1998) Effects of management on butterfly abundance in tallgrass prairie and pine barrens. Biol Conserv 83(1):77–89

Swengel AB (2001) A literature review of insect responses to fire, compared to other conservation managements of open habitat. Biodivers Conserv 10(7):1141–1169

Swengel AB, Swengel SR (2001) A ten-year study to monitor populations of the regal fritillary, *Speyeria idalia*, (Lepidoptera: Nymphalidae) in Wisconsin, U.S.A. Great Lakes Entomol 34:97–115

Swengel AB, Swengel SR (2007) Benefit of permanent non-fire refugia for Lepidoptera conservation in fire-managed sites. J Insect Conserv 11(3):263–279

Swengel SR, Schlicht D, Olsen F, Swengel AB (2011) Declines of prairie butterflies in the midwestern USA. J Insect Conserv 15:327–339

Taron DJ (1996) Insects. In: Packard S, Mutel C (eds) The tallgrass restoration handbook. Island Press, Washington, pp 305–318

Trager MD, Daniels JC (2009) Ant tending of Miami blue butterfly larvae (Lepidoptera: Lycaenidae): partner diversity and effects on larval performance. Fla Entomol 92:474–482

Vanreusel W, Van Dyck H (2007) When functional habitat does not match vegetation types: a resource-based approach to map butterfly habitat. Biol Conserv 135(2):202–211

Vogel JA, Debinski DM, Koford RR, Miller JR (2007) Butterfly responses to prairie restoration through fire and grazing. Biol Conserv 140(1):78–90

Vogel JA, Koford RR, Debinski DM (2010) Direct and indirect responses of tallgrass prairie butterflies to prescribed burning. J Insect Conserv 14(6):663–677

Wagner DL, Van Driesche RG (2010) Threats posed to rare or endangered insects by invasions of nonnative species. Ann Rev Entomol 55:547–568

The Imperiled Mardon Skipper Butterfly: An Initial Conservation Success

Rich Hatfield, Scott Hoffman Black, and Sarina Jepsen

Abstract The mardon skipper (*Polites mardon*), is a rare butterfly found only in the Pacific Northwest of the United States of America. The conservation of this butterfly has been part of an unprecedented cooperative effort between the Xerces Society for Invertebrate Conservation, The Forest Service/Bureau of Land Management Interagency Special Status/Sensitive Species Program, (ISSSSP) Washington Department of Fish and Wildlife, Washington State University, the Oregon Zoo, The Six Rivers National Forest, and the Center for Natural Lands Management.

In 2000 the mardon skipper was listed as a candidate species under the US Endangered Species Act. In an effort to conserve this butterfly partners developed and implemented surveys across the species' range, studied host plant preference and the impacts of management activities such as fire, and implemented restoration activities and management actions at dozens of mardon sites. This ongoing process identified many additional mardon skipper sites and an improved perspective on the status for the butterfly. The discovery of new sites was cited as one reason the U.S. Department of Interior, Fish and Wildlife Service did not ultimately list the species under the US Endangered Species Act. This process serves as a model that could be replicated toward the conservation of additional U.S. butterfly species, and potentially applied to additional animal groups.

Introduction

The mardon skipper (*Polites mardon*) (Fig. 1), is a rare butterfly found only in the Pacific Northwest of the United States of America. The conservation of this butterfly has been part of an unprecedented cooperative effort between the Xerces Society for Invertebrate Conservation, The Forest Service/Bureau of Land Management Interagency Special Status/Sensitive Species Program, (ISSSSP) Washington Department of Fish and Wildlife, Washington State University, the Oregon Zoo, The Six Rivers National Forest, and the Center for Natural Lands Management. In 2012, the mardon

R. Hatfield (✉) • S.H. Black • S. Jepsen
The Xerces Society for Invertebrate Conservation, Portland, OR, USA
e-mail: rich@xerces.org; sblack@xerces.org; sarina@xerces.org

© Springer Science+Business Media B.V. 2015

J.C. Daniels (ed.), *Butterfly Conservation in North America*,
DOI 10.1007/978-94-017-9852-5_7

Fig. 1 Mardon skipper (Polites mardon klamathensis Mattoon, J. Emmel and T. Emmel, 1998)

skipper working group was recognized for this conservation success through the Wings Across the Americas Butterfly Award.

In 2000 the mardon skipper was listed as a candidate species under the US Endangered Species Act. In an effort to conserve this butterfly partners developed *and* implemented surveys across the species' range, studied host plant preference and the impacts of management activities such as fire, and implemented restoration activities and management actions at dozens of specific mardon sites. This ongoing process identified many additional mardon skipper sites and an improved perspective on the status for the butterfly. The discovery of new sites was cited as one reason the U.S. Department of Interior, Fish and Wildlife Service did not ultimately list the species under the US Endangered Species Act. This process serves as a model that could be replicated toward the conservation of additional U.S. butterfly species, and potentially applied to additional animal groups. These efforts have led to a better conservation outlook for the mardon skipper but continued work is needed to ensure long-term viability.

History and Taxonomy of the Mardon Skipper

The mardon skipper was first described by W. H. Edwards (1881) from specimens taken near Tenino, Thurston County, Washington by H. K. Morrison (Dornfeld 1980). For nearly 100 years the only known populations were in Washington State.

In June 1979, Sterling and Eileen Mattoon discovered a population on High Divide Ridge in Del Norte County, California. Surveys in 2003 at the High Divide Ridge site (now called the Low Divide Road sites) identified four principal grasslands (totaling approximately 4.5 acres) in which mardon skippers were consistently observed (Haggard 2003). The first confirmed records for Oregon came in the late 1980s when John Hinchliff found three male specimens in collections held at the American Museum of Natural History in New York City. These specimens were collected by W.E. Lawrence in 1931 at Lake of the Woods, Klamath County. After Hinchliff's discovery, John Vernon and Mike Richard found a population on Dead Indian Road, Jackson County in 1990. Then, in 1991, Sterling Mattoon and others working on the Xerces Society Mount Ashland butterfly count added three additional populations in Jackson County, including the Soda Mountain Road population (now called the Hobart Peak complex) that eventually became the type specimen for the subspecies *Polites mardon klamathensis* (Mattoon et al. 1998).

There are two described subspecies of mardon skipper. Mattoon et al. (1998) proposed that the Oregon Cascades population be given subspecies status of *Polites mardon klamathensis*, and that the Washington and Northern California populations be given the subspecies status of *P. m. mardon*. Adults of *P. m. klamathensis* have a consistently tawnier dorsal and ventral coloration, when compared to adults from other populations (Warren 2005). A group of *P. mardon* populations exists in coastal Del Norte County approximately 80 air miles to the west of *P. m. klamathensis* populations. These were called *Polites mardon mardon* by Mattoon et al. (1998). However *P. mardon* from Del Norte County populations have not yet been carefully compared to *P. m. mardon* from Washington and the use of the name *P. m. mardon* for California populations should be considered tentative (Warren 2005).

Species Life History

Mardon skippers are part of the grass-feeding butterfly family, Hesperiidae, meaning the larvae feed exclusively on graminoids.

Mardon skippers are univoltine, completing one life cycle annually. Individual adults live between 5 days and 2 weeks. Adults do not all emerge on the same date, so the duration of the flight period at a given site depends, in part, on the population size and the environmental conditions. Sites with large populations may have a flight period that extends for more than a month, whereas sites with small populations may have adults present for only ten or fewer days. Weather influences emergence and flight period duration. Wet or cold conditions delay emergence and conversely, warm, dry conditions promote earlier emergence (Potter et al. 2002).

Adult mardon skippers eclose from late May to early July. Adults actively nectar throughout their flight period, and seek refuge from adverse weather low in the vegetative turf under grasses and forbs. *Polites mardon* seem to use a variety of flower species for nectar. The most common observations were on Vicia species (Washington sites) and *Potentilla diversifolia* Lehm. (Oregon sites) (Beyer and

Black 2007). Females lay eggs singly into graminoid bunches while perched (Beyer and Black 2007; Beyer and Schultz 2010). Eggs are thought to hatch within 7 days (Black and Vaughan 2005), and larvae feed through the late summer into the fall. Female mardon skippers have been observed ovipositing on multiple graminoid species, indicating that the larvae may be generalists (Beyer and Schultz 2010). Although mardon skippers do not select for specific graminoids during oviposition, they do exhibit plant specificity within localities. In the Washington Cascades, for example, the mardon skipper exhibits oviposition preferences for *Carex* spp. at one location and *Danthonia* spp. at another nearby location. Yet, when Idaho fescue (*Festuca idahoensis* Elmer) is present, it is generally preferred by ovipositing females (Beyer and Schultz 2010). In the Puget Sound prairies, the mardon skipper greatly prefers Roemer's fescue (*Festuca idahoensis* Elmer ssp. *roemeri* (Pavlick) S. Aiken) (Henry and Schultz 2012). At the two sites studied in Oregon, California oatgrass (*Danthonia californica* Bol.) was the most frequently utilized oviposition plant, although mardon skippers also oviposited on Roemer's fescue when available (Beyer and Black 2007). Variables such as graminoid structure, the surrounding plant structure, and tree shading also influence oviposition behavior (Henry and Schultz 2012). The degree to which these preferences are shown depends on the specific habitat available to each mardon skipper population (Beyer and Schultz 2010). Oviposition studies have not been conducted in California, but *Festuca idahoensis* is abundant at the known California mardon skipper sites (Haggard 2003).

It was previously believed that pupae hibernate through the winter (Potter et al. 1999; Dornfeld 1980; Newcomer 1966 in Potter et al. 1999), but preliminary studies of flagged mardon skipper larvae in the field by Beyer and Black (2007, see Fig. 2) suggest that Cascade populations overwinter in the larval life stage. In addition, in captive rearing studies, some mardon skippers appear to enter diapause and overwinter as larvae (Mary Jo Anderson, comm).

Occupancy by mardon skippers in meadow habitats is patchy; distribution is rarely homogeneous across an entire site (Beyer and Black 2007). This is likely dependent upon microclimatic conditions (Ehrlich 1992), and may be associated with palatability of the host plant due to proximity to a water source.

Status of the Mardon Skipper

The mardon skippers conservation status has been assessed as following: U.S. Federal Species of Concern, Washington State Endangered Species, Global Heritage Status Rank of G2G3, United States National Heritage Status Rank of N2N3, state rank of S1 in Washington and California, and S2 in Oregon, Forest Service Region 6 Sensitive Species, and a BLM Special Status Species. This butterfly was recently removed from the list of federal candidate species (USFWS 2012) partly because of the efforts of the Xerces Society and the Forest Service Interagency Special Status and Sensitive Species Program. Mardon skippers are grassland and open

Fig. 2 Mardon skipper (*Polites mardon klamathensis*) larva in fescue (Photograph by Loni Beyer)

meadow obligates endemic to the states of Washington, Oregon, and California. Historic mardon skipper ranges are not known, as historical documentation of this butterfly is scarce, with most surveys only occurring in the past decade (Black and Vaughan 2005). Despite the lack of data, mardon skippers are thought to have been more widespread and abundant prior to the past 150 years of human development, livestock grazing, fire suppression, and invasion of grassland habitat by native and non-native vegetation.

The mardon skipper is known from four geographic areas (Fig. 3):

1. southern Puget Sound in Washington
2. the east side of the Cascade Mountains in Washington
3. the Cascade Mountains in southern Oregon, and
4. Del Norte (north-coastal) California and the southern coast of Oregon

In 2010, the mardon skipper was known from 165 sites in Washington, Oregon, and California (Kerwin and Huff 2007; Department of the Interior, Fish and Wildlife Service 2012). Extensive surveys have been completed over the last decade to identify occupied habitats and those surveys have been widely successful, increasing the number of known occupied sites from 14 in 1999 to 165 in 2012. While it is likely that many of the occupied sites have been identified, there may be additional, undocumented sites, particularly in the Washington Cascades and in southwestern Oregon. However, despite the relatively large number of occupied sites, few studies have made efforts to determine the status or trends of mardon skipper populations at those sites. As such, the overall status of the mardon skipper remains relatively

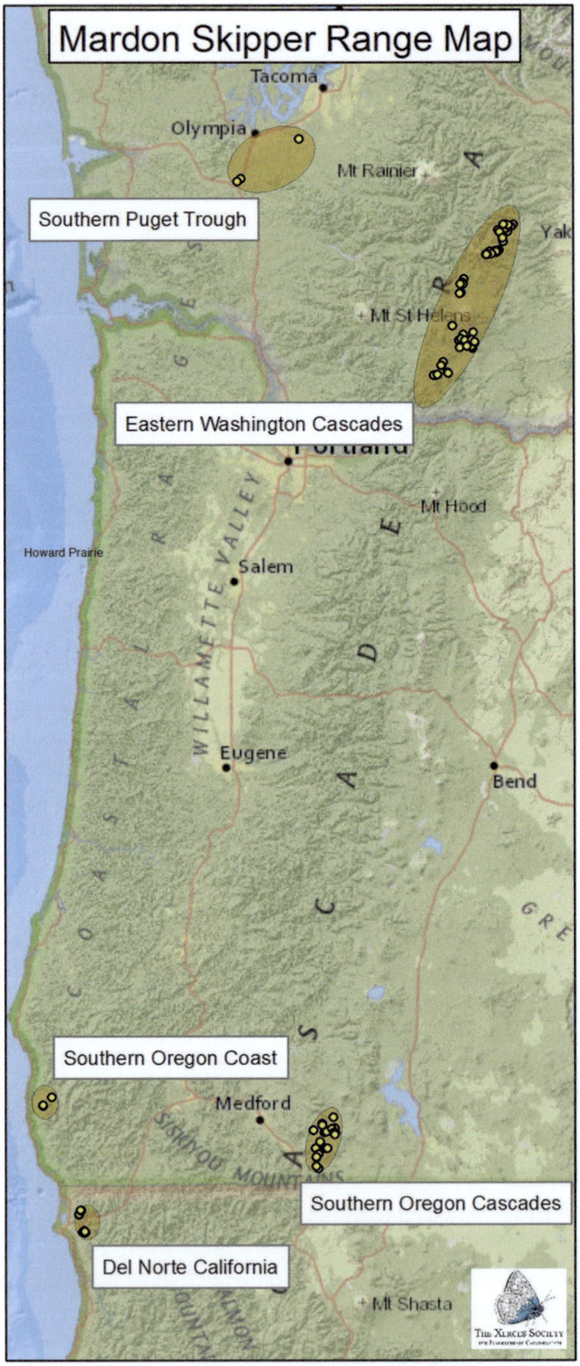

Fig. 3 Range-wide map of Mardon skipper distribution

unknown. Following are regional reports detailing the extent of what is known about mardon skipper status at the most well-studied locations. Most of the numbers reported are from 1-day counts and do not necessarily indicate population abundance or trends. However, beginning in 2013 with the implementation of the a range-wide monitoring protocol, several sites were selected for more intensive monitoring, with the ability to detect overall site abundance. There are several sites on private land, as well as a significant amount of habitat on the Yakama Nation lands in Washington for which very little information is available.

Washington

Status in the Southern Puget Sound Region (Adapted from Henry and Schultz 2012)

As recently as 1985 mardon skippers occurred on at least eight Puget prairie sites but are now present at only three (A. Potter, pers. comm., Fig. 4). Forest encroachment, urban and industrial development, and conversion to agriculture have reduced the

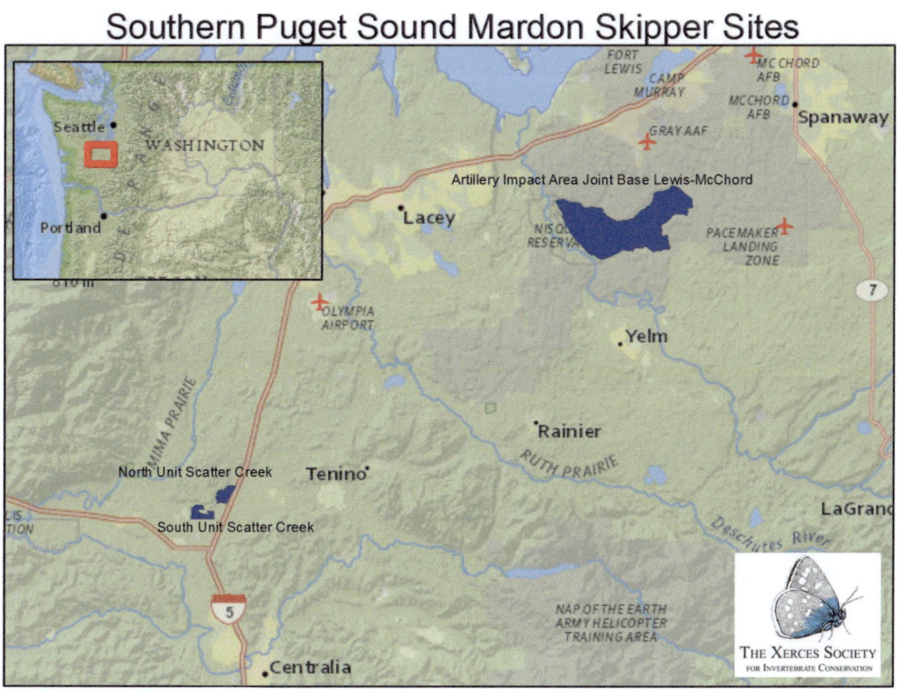

Fig. 4 Extant mardon skipper sites in the southern Puget Trough

Puget prairies to a fraction of their previous distribution (Crawford and Hall 1997). Historically, Native Americans maintained these prairies by setting frequent low-intensity fires (Storm and Shebitz 2006). In the absence of this disturbance regime, remaining prairie habitat has been degraded by the invasion of Scotch broom (*Cytisus scoparius* (L.) Link) and tall oatgrass (Arrhenatherum elatius (L.) P. Beauv. ex J. Presl and C. Presl). These species greatly alter the vegetation structure and outcompete the native grasses, Roemer's fescue (*Festuca idahoensis* ssp. *roemeri*) and California oatgrass (*Danthonia californica*), as well as a variety of native forbs.

There are ongoing monitoring and restoration efforts in the southern Puget trough, including a proposal to reintroduce mardon skippers to historic locations. Recent monitoring efforts in the southern Puget trough are among the most extensive surveys completed for the mardon skipper to date. In 2008–2009 Ann Potter, Insect Conservation Biologist with the Washington Department of Fish and Wildlife, led a team of investigators to complete a comprehensive distance sampling survey to estimate the mardon skipper population at two sites (Potter and Olson 2012). The population at each Scatter Creek site had peak counts in excess of 150 butterflies (Potter and Olson 2012). The population at the Lewis-McChord base has not been recently surveyed because of artillery fire (A. Potter, pers. comm.).

Eastern Cascades

Okanogan-Wenatchee National Forest In the Okanogan-Wenatchee National Forest, surveys between 2006 and 2012 located mardon skippers at 39 sites including 2 sites with 1-day counts with over 900 butterflies in 2008 and 2009. In 2012, three sites had 1-day counts with over 600 individuals (two of these three sites from 2012 are the same large sites as in 2008 and 2009). Work by the U.S. Forest Service, Washington State Department of Fish and Wildlife, the Xerces Society, and others have added to our knowledge of the species distribution on Okanogan-Wenatchee National Forest (Jepsen et al. 2007a, 2008; St. Hilaire et al. 2007, 2008, 2009, 2010). Since 2007, the Xerces Society has worked with the U.S. Forest Service to survey over 50 sites on the Naches Ranger District of the Okanogan-Wenatchee National Forest (see Fig. 5). New sites or populations have been documented during each year that surveys have been completed. Three new sites were discovered during the 2012 field season, and it is likely that additional undocumented sites exist on the Naches Ranger District. In 2014 one site on the Okanagan-Wenatchee National Forest was monitored using Distance Sampling. This site was visited on 5 separate occasions and a total of 1,482 butterflies were observed. The count on the observed peak of the flight season was 546 butterflies. Distance (Thomas et al. 2009) estimates the pooled daily population at this site to be 6,135 (15.84 % CV) mardon skippers (Fallon and Hatfield 2014).

Gifford Pinchot National Forest The mardon skipper is known from approximately 43 sites on the Gifford Pinchot National Forest (see Fig. 6) (Kerwin and Huff 2007). On the Gifford Pinchot National Forest, (elevation 549–1,677 m [1,800–

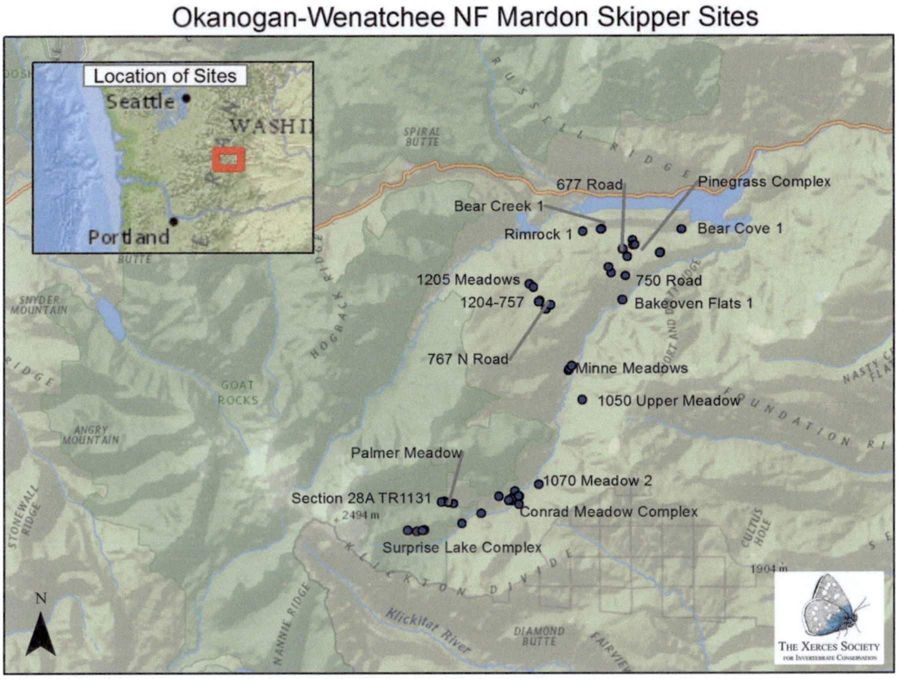

Fig. 5 Map of mardon skipper sites on the Okanogan-Wenatchee National Forest

5,500 ft]) the mardon skipper is found in warm, dry grand fir (*Abies grandis*) savannah/woodland with grassland intrusions or in small (<0.25 ha [<½ acre]) to larger grassland complexes. Site conditions range from dry, open ridge tops to areas associated with wetlands or riparian habitats (Potter et al. 1999, 2002). One-day population estimates on the Gifford Pinchot National Forest range from fewer than 10 individuals to greater than 350 individuals.

In 2014 one site complex on the Gifford Pinchot National Forest was monitored using Distance Sampling. This site was visited on 6 separate occasions and a total of 353 butterflies were observed. The count on the observed peak of the flight season was 85 butterflies. Distance (Thomas et al. 2009) estimates the pooled daily population at this site to be 2,687 (19.8 % CV) mardon skippers (Fallon and Hatfield 2014).

Yakama Reservation Since 2004, Yakama Nation biologists have conducted annual mardon skipper surveys within the Reservation. Mardon skipper populations have been detected in 11 populations on 23 different sites, many of which are extensive meadow complexes (Kerwin and Huff 2007).

Conboy Lake National Wildlife Refuge The mardon skipper is known from several locations on the Conboy NWR. Previous surveys were conducted on Conboy

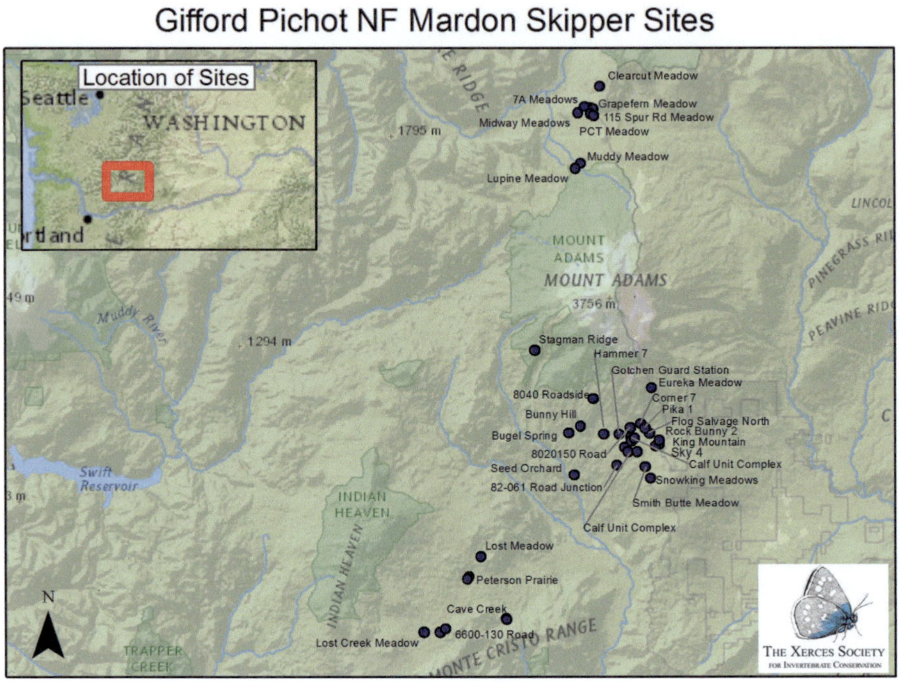

Fig. 6 Map of the mardon skipper sites on the Gifford Pinchot National Forest. Oregon

NWR from 1998 to 2001. The Xerces Society conducted additional surveys on the Refuge in 2013 to further determine the status and extent of mardon skipper populations. Over a 1 month period, Xerces staff visited the refuge a total of 7 days and documented several additional occupied sites on the Refuge. Sites that had previously been identified as occupied were also monitored using the range-wide monitoring protocol developed by the Xerces Society (Hatfield et al. 2013). Using distance sampling, pooled data for the two distance sampling units in the Laurel Unit reveal a density estimate of approximately 1,559 butterflies per hectare, with an estimated population of 8,734 butterflies. It is important to note that the survey area represents an area much smaller than the actual occupied habitat.

Oregon

Southern Cascades

All confirmed *P. m. klamathensis* populations occur in a small geographic area to the east of the city of Ashland in the Cascades of southern Oregon. All of the sites are encompassed in an area that is 18 miles from north to south and 8 miles from east

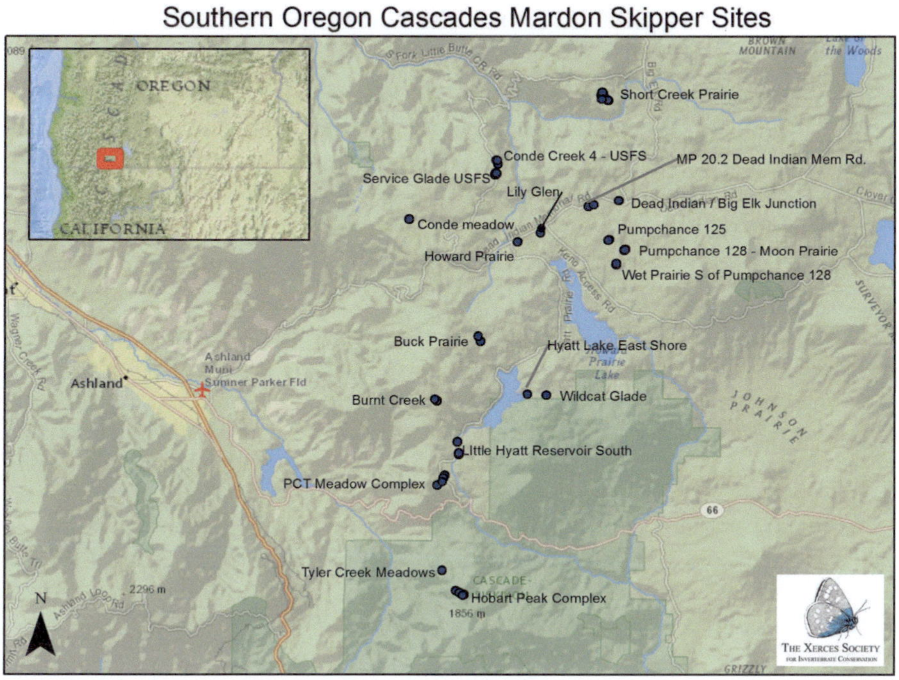

Fig. 7 Map of the mardon skipper sites in the southern Oregon Cascades

to west, with the southern border at Hobart Peak in the Cascade Siskiyou National Monument (see Fig. 7). The sites range in size from ¾ to 31 acres (1/3–13 ha) and are high-elevation (1,372–1,555 m [4,500–5,100 ft]) grasslands within mixed conifer forests that are associated with an ephemeral or permanent water source (Beyer and Black 2007). Work by the Xerces Society and others has added to our knowledge of the species distribution in southern Oregon (Vaughan and Black 2004, 2005; Black and Vaughan 2005; Ross 2005; Black 2006; Beyer and Black 2007; Jepsen et al. 2007b; Black et al. 2008a, b). Since 2004, the Xerces Society has worked with the U.S. Forest Service and the BLM to survey over 200 potential sites in the southern Oregon Cascades. With over 200 potential mardon skipper sites surveyed on public lands, we believe that the vast majority of sites on public lands in the southern Oregon Cascades have been located and that subsequent searches would yield few, if any, additional sites.

At present, 21 confirmed current and historic sites in 13 complexes are known from Forest Service and BLM lands in the Cascade Mountains in southwestern Oregon. Of these 21 sites on public lands, 17 were found to have extant populations of mardon skippers in surveys from 2009 to 2012. We have excluded some sites where in past years one individual was found in one survey and subsequent surveys yielded no positive findings, as these may be errant findings or vagrants and may not denote a site or population. Four known populations are located on private lands, although the status of those sites is not fully known.

Rogue River – Siskiyou National Forest Six sites are known from Rogue River – Siskiyou National Forest; two of these are in the Gold Beach Ranger District (GBRD) and four of them are located in the High Cascades Ranger District (HCRD). GBRD sites are included under a separate heading (Southern Oregon Coast). All four HCRD sites had extant populations of mardon skippers in 2011 and 2012.

BLM Sites There are 15 sites in 9 site-complexes on Bureau of Land Management (BLM) lands in the southern Oregon Cascades, which are all located in the Medford Management District. In 2012, no adult mardon skippers were observed on five of these sites (Conde Meadow, Wildcat Glade, Hobart Peak (East), PCT – North, and the meadow southwest of the PCT meadows), spanning four site complexes (Conde, Hobart Peak, Wildcat, and PCT). No mardon skippers were found on surveys of these sites in 2010 either and it is assumed that mardon have been extirpated from Wildcat glade and Conde meadows, but further surveys of Conde Meadow may be warranted as there is believed to be mardon skippers on private land close to the site. Especially with small population sizes, the accuracy of 1-day counts is highly dependent on conducting the count during the peak flight period, which varies from site to site based on microclimate factors. Notably, in 2010, no mardon skippers were located in the Hobart Peak (middle) site, but in 2012, we observed six mardon skippers at that site. The largest population of mardon skippers in this region was detected in 2012 at Howard Prairie. At this site in 2012, the largest single day count was 337 mardon skippers. This same site was monitored in 2014 using Distance Sampling. This site was visited on 7 separate occasions and a total of 169 butterflies were observed. The count on the observed peak of the flight season was 64 butterflies. Distance (Thomas et al. 2009) estimates the pooled daily population at this site to be 1,097 (26.4 % CV) mardon skippers (Fallon and Hatfield 2014).

Southern Oregon Coast

Coastal Oregon mardon skipper populations are generally associated with serpentine based soils that include plentiful bunchgrasses and varied nectar sources. Specimens of *Polites mardon* from coastal Oregon were first discovered at the McGuire Center for Lepidoptera and Biodiversity (Gainesville, Florida) by Andrew Warren with the label data "2 miles N of Gold Beach", suggesting the locality of collection to be the serpentine-bunchgrass hillsides to the north of the Rogue River near Wedderburn. In 2007, Dana Ross found a single male mardon skipper at the base of a bunchgrass hillside at Lone Ranch State Beach near Cape Ferrelo. The following year, in 2008, Dana Ross discovered a larger population at the Hunter Creek Area of Critical Environmental Concern (ACEC). In 2009, Dana Ross observed a single male in a small moist meadow at a new location on the Rogue-Siskiyou National Forest (Road #100 Meadow). In 2010, Dana Ross discovered two additional populations on the Rogue River – Siskiyou National Forest, one in the

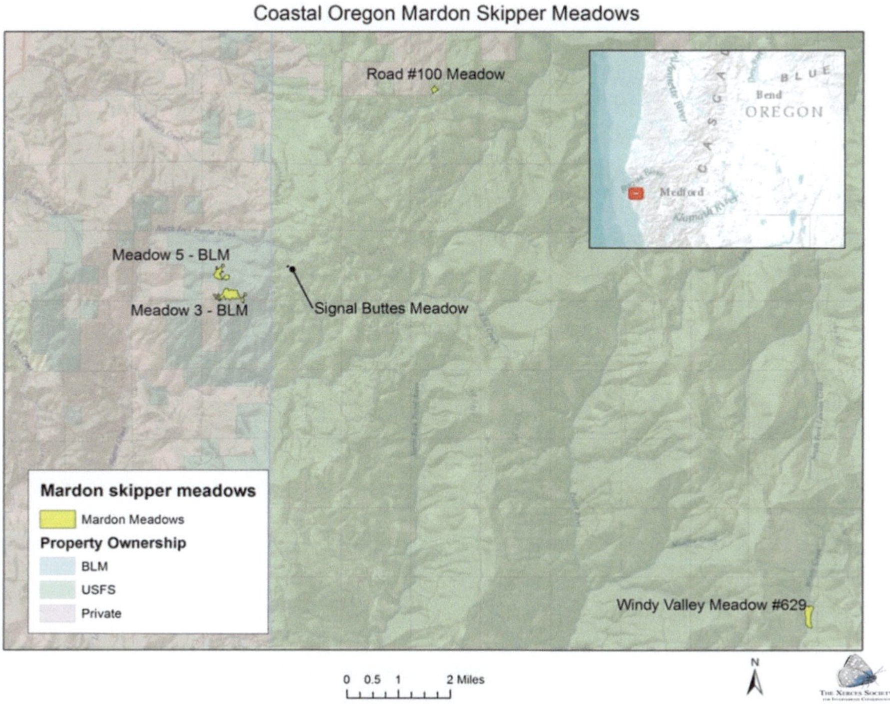

Fig. 8 Map of the mardon skipper sites in southwestern Oregon

Signal Buttes meadows adjacent to the Hunter Creek ACEC, and the other in the Windy Valley Meadow on the GBRD (see Fig. 8).

Aside from the single sighting near Cape Ferrelo, all confirmed coastal *Polites mardon* populations occur in a small geographic area to the east of the city of Gold Beach in the coast range of Southwest Oregon. All of the sites are encompassed in an area that is approximately 12 miles (N–S) by 12 miles (E–W) (see Fig. 8). All of the sites are relatively small (1–10 ha [2.5–25 ac]) and between the elevations of 595 and 900 m (1,952 and 2,952 ft).

Since most sites in this region seem to have low population sizes and have been found within the last 5 years, additional surveys are warranted. Dana Ross makes recommendations about potential habitat in his reports within the Rogue River – Siskiyou National Forest (Ross 2007, 2008, 2009, 2010). Additionally, Kip Wright (BLM wildlife biologist) mentioned that there are several lower elevation meadows within the Hunter Creek ACEC that may warrant searching (Kip Wright, pers. comm.).

Hunter Creek Areas of Critical Environmental Concern (ACEC) Two sites are known from the Hunter Creek ACEC, and both of these sites had extant populations during Xerces Society site visits in 2012.

Rogue River – Siskiyou National Forest Gold Beach Ranger District Three sites are known from the GBRD. All three sites have been discovered since 2009. Single day surveys suggest the Windy Valley Meadow had a relatively large population (>100 individuals). Additional surveys are needed at all three sites. Starting in 2012 mardon skippers at this site have been monitored several times per year. In 2012 the peak 1-day count was 54 individuals. In 2013, the peak 1-day count was 226 individuals. Starting in 2014, in consultation with the Xerces Society, agency biologists conducted distance sampling at Windy Valley. This site was visited on 7 separate occasions and a total of 275 butterflies were observed. The count on the observed peak of the flight season was 113 butterflies. Distance (Thomas et al. 2009) estimates the pooled daily population at this site to be 1,837 (14.02 % CV) mardon skippers (Fallon and Hatfield 2014). These estimates are likely low as the first day that mardon skippers were detected at the site was also the observed peak of the flight season.

California

Six Rivers National Forest

Surveys in 2003 at the High Divide Ridge site (now called the Low Divide Road sites) identified four principal grasslands (totaling approximately 4.5 acres (1.8 ha)) in which mardon skippers were consistently observed (Haggard 2003, Fig. 9). Three sites are in close proximity to each other (0.1 km), with the fourth being the most distant from these three (0.3 km). Dozens of individual mardon skippers have been detected during peak years in the 1–2 acre core area at this coastal California site (Haggard 2003). Surveys conducted by the Xerces Society in 2007 on 21 sites in the Six Rivers National Forest did not find additional populations. In 2008, Xerces Society staff re-surveyed several potential mardon sites on Rattlesnake Ridge and conducted new surveys of meadows at Coon Mountain (on the Six Rivers NF). Mardon skippers were discovered in the Coon Mountain meadow complex on June 10, 2008 (Fig. 9). This is believed to be the largest mardon skipper population in California based on a 1-day count of 204 individuals on June 10, 2008 (Black et al. 2008a) and follow up 1-day counts of 151 in 2009, 147 in 2010, 168 in 2011 and 313 in 2013. Since there are multiple years of multi-visit (2–3 per year) data at Coon Mountain, there is good baseline data to continue population monitoring at this site.

Redwood National and State Park (Adapted from 2011 RNSP Biological Assessment by Kristin Schmidt) There is one large complex known from Little Bald Hills in Redwood National and State Parks (Fig. 9). In 2009, eight monitoring transects were established throughout the National Park Service portion of the Little Bald Hills, in anticipation of prescribed burning or mechanical removal of conifers encroaching into the grasslands. Survey effort was approximately equal among the 2009–2011

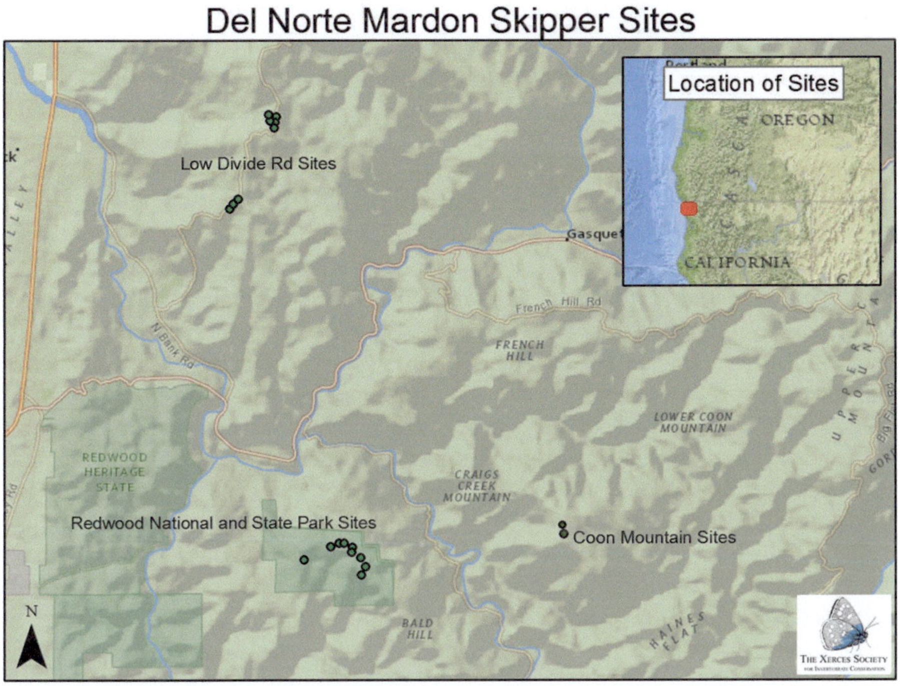

Fig. 9 Map of the mardon skipper sites in Del Norte County, California

surveys (n = 6 surveys in 2009 and 7 in 2010 and 2011). The maximum 1-day count in 2009 was 43, 27 in 2010, and 28 in 2011. Standard deviations were high due to surveys when no mardon skippers were detected. In 2009 and 2010, mardon skippers were first detected in mid-late May, and observations continued through late June or early July. In 2011 the first mardon skippers were observed in early June and continued to be detected through early July (RNSP, unpub. data). Using data from these transects, since the sites were visited 6–7 times per year, it would be possible to use INsect Count Analyzer (INCA, The Urban Wildlands Group 2002) to generate a baseline index of population size at these sites.

Threats and Current Management Practices

Mardon skipper populations face a variety of threats, including: grazing by domestic livestock; conifer, shrub, and grass encroachment; off road vehicle (ORV) use in meadows; prescribed and natural fire; recreation (including camping); applications of *Btk* (*Bacillus thuringiensis* var. *kurstaki*), climate change; and issues related to small population size and stochastic events.

Small and fragmented populations are generally at greater risk of extinction from normal population fluctuations due to predation, disease, and changing food supply, as well as from natural disasters such as droughts (reviewed in Shaffer 1981). Small populations are also threatened with extinction from a loss of genetic variability and reduced fitness due to the unavoidable inbreeding that occurs in such small populations (reviewed in Shaffer 1981).

Due to the local (cluster of habitat patches) and regional (four disjunct populations in WA, OR and CA) distribution of identified mardon sites, there is considerable evidence that the mardon skipper exhibits metapopulation dynamics. Metapopulations are defined as regionally distributed populations that are interdependent over time. There are many models of metapopulations, but a central theme to all of them is frequent local extirpations of individual habitat patches and subsequent recolonization, or rescuing, from nearby successful populations. In order for metapopulations to survive over time, there must be some opportunity for connectivity between patches of habitat. Considering changes in fire regime and land use, the connectivity of the landscape has changed significantly since 1950. A 1995 study noted a 39 % decrease in forest openings (meadow habitat) in the Klamath Mountains between 1944 and 1985, indicating significant forest encroachment in this time (Skinner 1995). Much of this forest encroachment happened along meadow connections, leading to further isolation of individual forest openings. Skinner (1995) also found that the median distance to the nearest forest break doubled between 1944 and 1985. A similar analysis in the Oregon Cascades found that the proportion of meadows occupied by trees increased from 8 to 35 % between 1950 and 2007 (Zald et al. 2012). From comparing photos from 1953 to 2011, similar forest encroachment has occurred throughout the mardon skipper range (see Fig. 10). Because of the limited dispersal abilities of mardon skippers, these changes have likely isolated remnant populations and reduced the probability of recolonization in the event of a local extirpation.

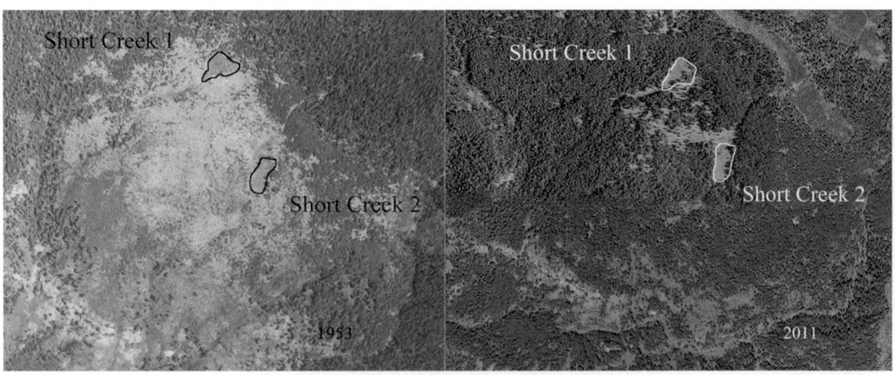

Fig. 10 Example of forest encroachment from aerial photographs of the Short Creek Complex on the Rogue River – Siskiyou National Forest in 1953 and 2011

Global climate change could threaten the mardon skippers' survival. Assessment of climate change trends in North America has already revealed changes in precipitation patterns, hydrology, and plant bloom time. Overall, annual mean air temperature increased in North America from 1955 to 2012. The effects of global climate change are projected to include warming in the western mountains, causing snowpack and ice to melt earlier in the season (Field et al. 2007), which will have an impact on mardon skipper sites as most are associated with permanent, ephemeral, or subsurface water. These climatic changes may lead to drier summer conditions, particularly in arid western areas where snowmelt is important to maintaining ephemeral water sources. Spring and summer snow cover has already been documented as decreasing in the western United States, and drought has become more frequent and intense (Parry et al. 2007; Saunders et al. 2008). Because the mardon skipper is associated with wet areas, its survival may be threatened by climate change induced habitat impairments. Moreover, since fewer seedlings and saplings will die in warmer winter and spring temperatures, and models predict an increase in fall precipitation, climate change seems to improve conditions for increased shrub and tree encroachment in open meadow habitats (Widenmaier and Strong 2010). Although management at the population scale cannot directly address global climate change, providing quality habitat spread in multiple sites across the landscape should help insulate species from climate change.

Below we provide general information on threats to mardon skippers from grazing (Fig. 11), conifer encroachment, prescribed fire, ORV use, invasive tall stature grass (i.e., meadow foxtail), and Btk applications. Mardon skippers persist in open prairie and meadow ecosystems. These ecosystems are naturally maintained

Fig. 11 Grazing impacts at Forest Service Conde 4. Stubble height was very low with some areas reduced to bare soil (Photo by Scott Hoffman Black)

by disturbance (Coop and Givnish 2007). Yet, a delicate balance is needed to create sufficient disturbance to restore and maintain the system without creating so much disturbance that the habitat is further degraded (Schultz and Crone 2008).

Livestock Grazing

Graminoid cover, composition, and structure are important for mardon skippers (Beyer and Schultz 2010) because they provide larvae with adequate food, microclimate, and protection from predation and parasitism. Livestock grazing can adversely impact butterfly populations directly by trampling during immobile life stages (egg, larvae, pupae) or during cool temperatures when adult movement is restricted (Warren 1993). Grazing can also be detrimental to butterfly populations indirectly by altering plant community composition (Stoner and Joern 2004) and stripping habitat of vegetation, removing adult nectar resources, introducing invasive weeds (Hayes and Holl 2003), and changing meadow hydrology (Belsky et al. 1999). In a study on the Dakota skipper (*Hesperia dacotae*) Dana (1997) found that grazing cattle reduced skipper numbers in direct proportion to grazing intensity. However, light rotational grazing can maintain vegetation heights and habitat heterogeneity favorable to some butterflies (Ravenscroft 1994). For instance, the silver-spotted skipper (*Epargyreus clarus*) has not only shown a positive response to moderate grazing, but depends on it to maintain the structure of its host plant (Thomas and Jones 1993; Davies et al. 2005). In order for grazing to have a minimal impact on the mardon skipper, the combination of timing, intensity, and duration must be found that is best suited for the specific ecology of the butterfly. Runquist (2011), in a study focusing specifically on the butterfly taxa of the Cascade-Siskiyou National Monument, found that graminoid-feeding butterflies were 1/3 less abundant at higher grazing utilization classes (>30 % utilization). In some sites the correct solution may be to eliminate grazing, while in other sites some grazing at the appropriate time of the year may not negatively impact the site. The timing and frequency of livestock rotation depends on both the size of the herd and the size of the grazed area (Schtickzelle et al. 2007). Generally speaking, grazing periods should be short with relatively long recovery periods for the habitat, and grazing should not take place during the adult flight period since butterflies are utilizing nectar sources and looking for oviposition sites during this time.

To date, efforts to manage the effects of grazing on mardon skippers have been varied. At several sites in the southern Oregon Cascades, temporary cattle exclosures have been constructed around key habitat to help exclude cattle during the adult flight season. Similar efforts to control grazing have been constructed on the Okanogan-Wenatchee National Forest in southern Washington. The efficacy of these exclosures on improving habitat for mardon skippers or mardon skipper populations has not been closely monitored, and would be worthy of more attention from land managers.

Forest Encroachment

The removal and alteration of the natural disturbance regimes (such as fire suppression), that once maintained low shrub and conifer seedling establishment rates, has led to the loss and degradation of forest-meadow ecosystems (Coop and Givnish 2007; Norman and Taylor 2005). Forest encroachment not only reduces the amount of open habitat, but closes off corridors between meadows, reducing butterfly dispersal (Roland and Matter 2007). During the adult flight period, mardon skippers avoid heavily forested habitats, avoid forest edges and trees during oviposition, and are assumed to have limited dispersal abilities (Runquist 2004; Beyer and Black 2007; Beyer and Schultz 2010). Large dense shrubs likely have a similar adverse impact as encroaching trees to the habitat and behaviors of this butterfly.

The lack of genetic interchange and movement of individuals between populations will likely lead to lower overall fitness and extirpations of local populations. A better understanding of the feasibility and the effects of reconnecting neighboring mardon skipper populations that have been isolated by forest and shrub encroachment is warranted (Bergman 1999; Dennis et al. 2006).

Forest encroachment is a pervasive threat throughout many mardon skipper sites and meadows throughout the Pacific Northwest. As a result of conservation efforts by ISSSSP and the Xerces Society, site management plans have been written and implemented for many mardon skipper sites which identify this threat. As a result, encroaching conifers have been removed from many of the most well-known sites helping to secure over 100 ha of habitat for mardon skippers and other open meadow obligates.

Non-native Vegetation

Changes of vegetation structure at oviposition sites can impact larval survival and host plant selection by ovipositing mardon skippers (Henry and Schultz 2012). Mardon skippers prefer open habitat with isolated patches of host plant for oviposition (Beyer and Schultz 2010; Henry and Schultz 2012). The invasion of tall-stature grasses can dramatically change the structure of the vegetation and therefore the microhabitats within the meadow. Runquist (2011) believes that tall-stature grasses have led to a decline in some mardon populations on BLM lands. Scotch broom (*Cytisus scoparius*) and tall oatgrass (*Arrhenatherum elatius*) is also implicated in altering mardon skipper habitat in Southern Puget Sound sites (Potter and Olson 2012). Butterfly populations are constrained by temperature (Crozier 2003, 2004) and are restricted to especially warm macro- and microhabitats in temperate climates (Thomas et al. 2001). The small tufts and open habitat structure selected by ovipositing females are likely to correspond to the warmest locations in the prairie or meadow (Forsberg 1987; Stoutjesdijk and Barkman 1992). The invasion of a tall-stature grass, such as meadow foxtail (*Alopecurus pratensis*) or tall oatgrass, will significantly change the sun exposure and moisture content of invaded areas, thereby changing the attractiveness to mardon skippers and survival of larvae, even if desired host plants remain in the understory of the tall-stature grass.

Prescribed Fire

Due to the importance of fire in maintaining many native ecosystems, the use of controlled burning as a management tool has become increasingly common. The effects of fire on vegetation and vertebrate communities are more widely understood than the effects of fire on invertebrates. Fire can benefit, harm, or have no significant effect on invertebrates depending on the biology of the specific taxa (Gibson et al. 1992) and the specifics of the burn. Burning of meadows with populations of butterflies, such as the mardon skipper, could extirpate the population if not done with careful consideration of butterfly behavior and life stage at time of burn, and knowledge of where the skipper population is distributed across the meadow. In addition, with isolated populations, there are often no source populations available for re-colonization once a population has been locally extirpated.

A recent study on prescribed burning and the mardon skipper in California showed substantially fewer butterflies in the burned areas of meadows compared to unburned areas after 1, 2, and 3 and 5 years following the burn event (Black et al. 2014). Counts across all survey dates in 2009, 2010, 2011 and 2013 showed mardon numbers that ranged from 1.7 to 27 times higher in unburned zones compared to burned zones on the same dates. When we pool transect data within years, both burning and time have a significant effect on mardon skipper abundance. The fact that there is no interaction effect between time and burning suggests that the effect of burning on mardon skippers is real and not confounded by annual variation in butterfly populations (Black et al. 2014). One thing worth noting is that mardon skipper populations in burned areas do appear to be making a comeback after 5 years. This accentuates the need to leave substantial habitat when using fire as a management tool for mardon skippers. Leaving enough habitat so that a large enough butterfly population remains to sustain a prolonged (although ideally short-term) decrease in habitat quality is essential.

Off Road Vehicles

Mardon skipper eggs, caterpillars, pupae, and adults are killed by recreational activities such as off-road vehicle (ORV) driving (Potter et al. 1999). These activities can also damage native host plants and may lead to an invasion of non-native plant species. ORVs aid the dispersal of invasive non-native species deep into forestlands. One study found that in just one trip on a 10 mile course, an ORV dispersed 2,000 spotted knapweed seeds (Trunkle and Fay 1991). In Wisconsin, a survey of seven invasive plant species along ORV routes found at least one of these (exotic) plant species on 88 % of segments examined (Rooney 2005).

Throughout this species' range, significant efforts have been made to curb the use of ORVs in mardon skipper habitat. These projects have involved road closures, installation of physical barriers such as boulders and fences, and the installation of educational signage. As a result of these projects and adaptive management, the threat of ORVs has been removed from many sites where it used to be a significant risk to mardon skippers and their habitat.

Use of Btk

Btk (*Bacillus thuringiensis* var. *kurstaki*), a Lepidoptera-specific insecticide, has been widely used to treat insect pests in western forests (Wagner and Miller 1995). Btk is a bacterium which, when ingested, is lethal to butterfly and moth larvae.

Species such as the mardon skipper that are single-brooded with spring-active larvae, which feed during the application period for the target species are especially vulnerable to Btk (Wagner and Miller 1995). Because of the mardon skipper's current patchy distribution on isolated sites and low vagility, its populations are even more threatened by Btk applications due to the decreased probability of re-colonization.

The threat of Btk is heightened because Btk has been shown to drift at toxic concentrations for distances greater than 2 miles from target spray areas (Barry et al. 1993; Whaley et al. 1998). As a result, aerial spraying with Btk can have significant adverse effects on mardon skippers in the general area of an aerial Btk spray project.

Past Monitoring Efforts and Analysis

Prior to 2013 most monitoring efforts for mardon skipper populations had been inadequate for any population level analysis. Most surveys that were conducted have taken place on a single day, ideally near the peak of flight season, but were not conducted consistently between years. The current database has the power to detect the minimum population size per site and presence/non-presence data. While these data are important, they are insufficient for effective and evidence based conservation efforts. In order to have the power to achieve population level analysis, it is essential, to survey a population multiple times during its flight season at multiple sites.

In addition to 1-day counts, there have been extensive efforts to survey potential habitat for new populations. These efforts have been widely successful, and are largely responsible for the change in status of this butterfly from warranted to delisted by the U.S. Fish and Wildlife Service (USFWS) in September of 2012 (Department of the Interior, Fish and Wildlife Service 2012).

Efficacy of Current Management Practices

Due to the lack of population level data on the mardon skipper (see "Past Monitoring Efforts and Analysis" above), the quantitative efficacy of current management practices is unknown. We knew very little about the population status of the mardon skipper before management was implemented on sites throughout the butterfly's range, and monitoring efforts have not changed after implementation. Because of this, we do not know to what degree any potential threats have been mediated by conservation efforts.

Despite the fact that there is no direct evidence for efficacy of management activities to date, some activities have very likely improved conditions for mardon skippers. Mardon skippers are open meadow habitat specialists and all efforts to maintain this habitat has likely been beneficial. As such, the significant efforts to curb forest encroachment have likely had a significant role in the conservation of this species. Similarly, removing ORVs in open meadow habitat has helped to preserve meadow habitat by slowing the spread of invasive species while preventing the direct trampling of mardon skippers or their habitat.

The management recommendations regarding cattle grazing (creating cattle exclosures, altering allotment start dates, etc.) have been somewhat experimental, yet, unfortunately, no effort has been made to detect the effects of these treatments. As a result, we have no empirical data about the degree to which grazing poses a threat to mardon skippers or their habitat, or how efforts to alter grazing intensity affect mardon skipper populations. There is little doubt that grazing poses both indirect (changes in vegetation structure, composition, and phenology) and direct (trampling) threats to mardon skippers. It is also known that mardon skippers have coexisted with livestock since livestock grazing was introduced in the mid to late nineteenth century. So while grazing may pose a threat to mardon skippers, particularly at higher stock rates and utilization levels, the degree of the threat is unknown. It is also possible that current threats due to livestock grazing are greater than historical threats due to the cumulative effects of over a century of livestock grazing on grasslands (Fleischner 2002).

The most thorough review of the effects of management on the mardon skipper has taken place on Coon Mountain in Del Norte California, where the effects of a 2008 prescribed fire have been measured on mardon skippers. Even though the burn was not planned for the purpose of mardon skipper management, it presented an opportunity to monitor the effects of controlled burns on mardon skippers (Black et al. 2009). This study showed substantially fewer butterflies in the burned areas of meadows compared to unburned areas after 1, 2, and 3 and 5 years following the burn event (Black et al. 2014). Counts across all survey dates in 2009, 2010, 2011 and 2013 showed mardon numbers that ranged from 1.7 to 27 times higher in unburned zones compared to burned zones on the same dates. When we pool transect data within years, both burning and time have a significant effect on mardon skipper abundance. The fact that there is no interaction effect between time and burning suggests that the effect of burning on mardon skippers is real and not confounded by annual variation in butterfly populations (Black et al. 2014).

Captive Propagation

Captive rearing and reintroduction is likely an essential component in the recovery of mardon skipper in the Puget Trough where population numbers are very low. Unfortunately stimulation of egg laying, asynchrony of larval development, and pupal eclosion timing have proven to be difficulties in mardon skipper captive rearing (Grosboll 2004). Therefore efforts to develop mardon skipper rearing methods that could be used to produce large numbers of skippers in captivity have not been successful to date (Linders 2007).

Future Conservation

When attention was first drawn to this species in the late 1990s and early 2000s, the mardon skipper was known from only ten sites. Since then, because of extensive searches by USFS, BLM, Washington Department of Fish and Wildlife, the Xerces Society, and others, the number of known sites with mardon skippers has expanded to over 165. However, while the number of sites has increased dramatically, we still know very little about the population trend at any of these individual sites, let alone the overall population. We also know very little about how these sites are interacting with each other (i.e., are they behaving as metapopulations?), and more information about dispersal would help inform conservation efforts. It is certainly true that the increased number of known sites is a positive sign for the long-term conservation of this species, as a local extirpation will not have as negative an effect on the overall population. However, it is also possible that due to forest encroachment and habitat fragmentation that each of these sites is more isolated than they were historically and therefore more at risk from extirpation without chance of re-colonization.

A recent meta-analysis done by Thomas et al. (2011) determined that the main determining factor of butterfly abundance in almost all butterfly conservation studies is the quality of larval host plant habitat. In most cases, it is not merely the abundance of the host plant that matters, but the quality (size, degree of isolation) of the host plant as determined by oviposition (Thomas et al. 2011). This agrees with what we have learned from Beyer and Schultz (2010) and Henry and Schultz (2012), which determined that the mardon skipper prefers isolated patches of its host plant surrounded by bare ground or moss. Notably, mardon skippers actively avoided host plants that were in the understory of taller stature grasses. Because of the work that has recently been conducted on mardon skipper host plant preference, it should be fairly easy to detect suitable patches for the mardon skipper in occupied sites, at least in regions in which host plant preference has been studied. What we do not know is what level of disturbance or management helps to maintain high quality habitat. Making an effort to test the effects of different management alternatives on mardon skipper habitat (e.g., host plant condition) should be of high priority.

However, in order to test the effect of management activities on mardon skipper populations, the new monitoring protocol needs more broadly implemented.

Currently, the only data that exists for mardon skipper populations are 1-day counts that were taken "near the peak flight period" of the butterfly. Unfortunately we have no idea how close these counts were to the peak flight period, and therefore do not have a reasonable estimate of the minimum population that exists at each surveyed site. These data, while useful for detecting presence, and potentially identifying local extinctions, do not provide enough detail to detect changes in population size.

Key Remaining Questions

- What is the size and trend of mardon skipper populations at individual sites?
- What is the size and trend of mardon skipper populations in individual regions?
- What is the size and trend of mardon skipper populations range-wide?
- What are the direct and/or indirect effects of grazing on mardon skipper populations?
- Are mardon skippers from each of the regions genetically distinct (i.e. separate subspecies)—and if so, how should we prioritize the conservation of each genetically distinct unit?
- What management activities significantly increase mardon skipper populations and improve mardon skipper habitat?
- Are threats more significant for adults, eggs, or larvae?
- What are the impediments to effective captive propagation?

Priorities for Future Conservation

1. Maintain conservation efforts for the species at existing sites. Efforts should be made to ensure that sites are protected from overgrazing, off road vehicles, conifer encroachment and invasive species encroachment.
2. Implement the stratified monitoring scheme developed by the Xerces Society to determine population level data for each region of the mardon skipper.
   (a) Priorities should be placed on determining daily and total population sizes, along with reasonable detectability statistics.
   (b) At a minimum, a concerted effort to determine an estimate of the population growth rates (μ) and the variation in growth over time (σ^2) must be made. Without these data, monitoring the effects of management treatments on butterflies, or even detecting changes in population size, will be impossible.
3. Determine the effects of livestock grazing on mardon skipper populations and their habitat.
   (a) Priorities should be placed on determining how stocking rates and utilization levels affect mardon skipper populations and host plant condition.

   (b) An additional priority should be placed on determining the effects of cattle exclosures on mardon skipper populations and habitat.
   (c) These studies should be repeated in all regions where livestock grazing is a factor as host plant selection differs across the range and the results from one region may not be applicable regionwide.

4. Conduct genetic analyses to determine the subspecies status of all four regions of mardon skipper populations. The Washington populations are currently considered to be the same subspecies as the Del Norte and Oregon Coast populations (Mattoon et al. 1998; Warren 2005; Pelham 2008). However, given the geographic separation between all four populations, there is certainly the possibility that there are more than two subspecies regionwide. Efforts to determine subspecies status of all four regions would greatly enhance conservation efforts and help to determine regional priorities.
5. Work to understand stimulation of egg laying, asynchrony of larval development, and pupal eclosion timing so that a captive propagation program can be developed for augmentation and/or reintroductions into Puget Trough sites.

Summary

Over the last 15 years a tremendous amount of work and effort has been directed towards the conservation of the mardon skipper butterfly. In that timeframe a number of high priority goals have been achieved and a tremendous amount of progress has been made. We have conserved many sites and restored others. We have also ameliorated many threats, especially on sites managed by the US Forest Service and BLM. We have a much better understanding of the lifecycle of the butterfly, we have an expanded understanding of the mardon skippers' range, and conservation plans have been implemented to improve the meadow habitat upon which mardon skippers depend.

Even though substantial progress has been made, there is still much to do to ensure we understand the conservation status of this butterfly and ensure that it is protected into the future.

Acknowledgements We would like to acknowledge all of the biologists, and butterfly enthusiasts that have contributed to the extensive conservation efforts for this butterfly.

References

Barry JW, Skyler PJ, Teske ME et al (1993) Predicting and measuring drift of Bacillus thuringiensis sprays. Environ Toxicol Chem 12:1977–1989

Belsky A, Matzke A, Uselman S (1999) Survey of livestock influences on stream and riparian ecosystems in the western United States. J Soil Water Conserv 54:419–431

Bergman KO (1999) Habitat utilization by Lopinga achine (Nymphalidae: Satyrinae) larvae and ovipositing females: implications for conservation. Biol Conserv 88:69–74

Beyer L, Black SH (2007) Site utilization by adults and larvae of Mardon skipper Butterfly (Polites mardon) at four sites in Washington and Oregon

Beyer LJ, Schultz CB (2010) Oviposition selection by a rare grass skipper Polites mardon in montane habitats: advancing ecological understanding to develop conservation strategies. Biol Conserv 143:862–872

Black SH (2006) Report to U.S. Forest Service: Xerces Society Surveys for Polites mardon in the Rogue River National Forest in Southern Oregon. The Xerces Society, Portland, Summer 2006

Black S, Vaughan D (2005) Species profile: Polites mardon. http://www.xerces.org/mardon-skipper/

Black SH, Jepsen S, Lauvray L (2008a) Report to the U.S. Fish and Wildlife Service, Surveys for Polites mardon in Del Norte County of Northern California. The Xerces Society, Portland

Black SH, Jepsen S, Lauvray L (2008b) Report: Xerces Society Polites mardon surveys in southern Oregon's Mt. Ashland area. The Xerces Society, Portland, Summer 2008

Black S, Mazzacano C, Lauvray L (2009) Mardon skipper survey of Coon Mountain burn site. Report to the U.S. Fish and Wildlife Service and the U.S. Forest Service, p 17

Black S, Fallon C, Hatfield R, Mazzacano C (2014) Controlled burning and Mardon skipper: summary of Mardon skipper Coon Mountain burn site occupancy study data from 2009 to 2013. Report to the U.S. Forest Service, Oregon Zoo, and U.S. Fish and Wildlife Service:29

Coop JD, Givnish TJ (2007) Spatial and temporal patterns of recent forest encroachment in montane grasslands of the Valles Caldera, New Mexico, USA. J Biogeogr 34:914–927

Crawford RC, Hall H (1997) Changes in the south Puget prairie landscape. Dunn, PV, Ewing K. Retrieved 2 Jan 2013, from http://ww.southsoundprairies.org/tech/Changes%20in%20SPuget%20Sound%20Landscape.pdf

Crozier L (2003) Winter warming facilitates range expansion: cold tolerance of the butterfly Atalopedes campestris. Oecologia 135:648–656

Crozier L (2004) Warmer winters drive butterfly range expansion by increasing survivorship. Ecology 85:231–241

Dana R (1997) Characterization of three Dakota skipper sites in Minnesota. Unpublished report, Minnesota Department of Natural Resources, Natural Heritage and Nongame Research Program, St. Paul 22. Retrieved 2 Jan 2013, from http://files.dnr.state.mn.us/eco/nongame/projects/consgrant_reports/1997/1997_dana.pdf

Davies ZG, Wilson RJ, Brereton TM, Thomas CD (2005) The re-expansion and improving status of the silver-spotted skipper butterfly (Hesperia comma) in Britain: a metapopulation success story. Biol Conserv 124:189–198

Dennis RLH, Shreeve TG, Van Dyck H (2006) Habitats and resources: the need for a resource-based definition to conserve butterflies. Biodivers Conserv 15:1943–1966

Department of the Interior, Fish and Wildlife Service (2012) Endangered and threatened wildlife and plants; 12-month finding on a petition to list the Mardon skipper as threatened or endangered; proposed rule, pp 54332–54352. Federal Register

Dornfeld J (1980) Butterflies of Oregon. Timber Press, Forest Grove

Ehrlich PR (1992) Population biology of checkerspot butterflies and the preservation of global biodiversity. Oikos 63:6–12

Fallon C, Hatfield RG (2014) Mardon skipper (*Polites mardon*) distance sampling surveys at four sentinel sites in Oregon and Washington. Final report to the U.S. Forest Service, Bureau of Land Management, and the Interagency Special Status/Sensitive Species Program (ISSSSP). The Xerces Society, Portland

Field CB, Lobell DB, Peters HA, Chiariello NR (2007) Feedbacks of terrestrial ecosystems to climate change. Annu Rev Environ Resour 32:1–29

Fleischner TL (2002) Ecological costs of livestock grazing in western North America. Conserv Biol 8:629–644

Forsberg J (1987) Size discrimination among conspecific hostplants in two pierid butterflies; Pieris napi L. and Pontia daplidice L. Oecologia 72:52–57

Gibson CWD, Brown VK, Losito L, McGavin GC (1992) The response of invertebrate assemblies to grazing. Ecography 15:166–176

Grosboll D (2004) Captive rearing the endangered Mardon skipper (Polites mardon) and Taylor's checkerspot (Euphydryas editha taylori) butterflies: initial results (Lepidoptera, Nymphalidae). In: Hooper TD (ed) Proceedings of the species at risk, pathways to recovery conference, Victoria, 2004. Species at Risk Pathways to Recovery Conference Organizing Committee, Victoria, 1–6 March 2004

Haggard J (2003) 2003 status report for the mardon skipper (Polites mardon mardon) in Del Norte County, p 3

Hatfield R, Black SH, Jepsen, S (2013) Range-wide strategy and monitoring protocol for the mardon skipper (*Polites mardon*, W.H. Edwards, 1881). The Xerces Society for Invertebrate Conservation. Portland, OR

Hayes GF, Holl KD (2003) Cattle grazing impacts on annual forbs and vegetation composition of mesic grasslands in California. Conserv Biol: J Soc Conserv Biol 17:1694–1702. Blackwell Science Inc

Henry EH, Schultz CB (2012) A first step towards successful conservation: understanding local oviposition site selection of an imperiled butterfly, mardon skipper. J Insect Conserv 17:183–194

Jepsen S, Black SH, Lauvray L (2007a) Xerces Society surveys for Polites mardon mardon in the Naches Ranger District of Washington. The Xerces Society, Portland, Summer 2007

Jepsen S, Black SH, Lauvray L (2007b) Report: Xerces Society Polites mardon surveys in southern Oregon. The Xerces Society, Portland, Summer 2007

Jepsen S, Black SH, Lauvray L (2008) Xerces Society surveys for Polites mardon mardon in the Naches Ranger District, Wenatchee National Forest of Washington. The Xerces Society, Portland, Summer 2008

Kerwin AE, Huff R (2007) Conservation assessment for the Mardon skipper (Polites mardon). Version 2.0. December 2011. USDA forest service region 6. USDI Bureau of Land Management. Oregon/Washington, DC, 42 pp

Linders M (2007) Captive rearing methods test for mardon skipper (Polites mardon). Report to the Nature Conservation/Army Compatible Use Buffer Program at Joint Base Lewis McChord

Mattoon S, Emmel J, Emmel T (1998) The distribution of Polites mardon (Lepidoptera: Hesperiidae) in North America, and description of a new subspecies from southern Oregon. Systematics of western North American butterflies, vol 878. Mariposa Press, Gainesville, pp 767–774

Newcomer E (1966) Life histories of three western species of Polites. J Res Lepid 5:243–247

Norman SP, Taylor AH (2005) Pine forest expansion along a forest-meadow ecotone in northeastern California, USA. For Ecol Manage 215:51–68

Parry ML, Canziani OF, Palutikof JP et al (2007) IPCC, 2007: climate change 2007: impacts, adaptation and vulnerability. Contribution of working group II to the fourth assessment report of the intergovernmental panel on climate change. Cambridge University Press, Cambridge

Pelham JP (2008) A catalogue of the butterflies of the United States and Canada: with a complete bibliography of the descriptive and systematic literature. Lepidoptera Research Foundation

Potter A, Olson G (2012) Monitoring mardon skipper (Polites mardon) on the Scatter Creek Wildlife Area. Washington Department of Fish and Wildlife. Olympia, WA

Potter A, Fleckenstein J, Richardson S, Hays D (1999) Washington state status report for the mardon skipper. Washington Department of Fish and Wildlife, Olympia

Potter A, Fleckenstein J, Feen J (2002) Mardon skipper range and distribution in Washington in relation to state and federal highways with a habitat description and survey method guidelines. Final report to Washington Department of Transportation. Olympia, WA

Ravenscroft NOM (1994) The ecology of the chequered skipper butterfly Carterocephalus palaemon in Scotland. I. Microhabitat. J Appl Ecol 613–622

Roland J, Matter SF (2007) Encroaching forests decouple alpine butterfly population dynamics. Proc Natl Acad Sci 104:13702–13704

Rooney TP (2005) Distribution of ecologically-invasive plants along off-road vehicle trails in the Chequamegon National Forest, Wisconsin. Mich Bot 44:178–182

Ross E (2005) Surveys for Mardon skipper (Polites mardon klamathensis) on public lands in southern Oregon. A report submitted to the US Fish and Wildlife Service, Portland

Ross D (2007) Surveys for Mardon skipper (Polites mardon) in southwestern Oregon. A report submitted to the Xerces Society for Invertebrate Conservation and the US Fish and Wildlife Service, Portland

Ross D (2008) Surveys for Mardon skipper (Polites mardon) in coastal southern Oregon

Ross D (2009) 2009 surveys for Mardon Skipper (Polites mardon) on the Gold Beach Ranger District of the Rogue River-Siskiyou National Forest. Corvallis, OR

Ross D (2010) 2010 surveys for Mardon Skipper (Polites mardon) on the Rogue River National Forest. Corvallis, OR

Runquist E (2004) Line transect sampling and surveys for Mardon skippers (Polites mardon) in Del Norte County, California. Report submitted to USFWS, Arcata

Runquist E (2011) Butterflies, cattle grazing, and environmental heterogeneity in a complex landscape. J Res Lepid 44:61–76

Saunders S, Montgomery C, Easley T (2008) Hotter and drier: the West's changed climate. Prepared for The Rocky Mountain Climate Organization and Natural Resources Defense Council, p 64. Retrieved 2 Jan 2013, from http://www.nrdc.org/globalWarming/west/west.pdf

Schtickzelle N, Turlure C, Baguette M (2007) Grazing management impacts on the viability of the threatened bog fritillary butterfly Proclossiana eunomia. Biol Conserv 136:651–660

Schultz CB, Crone EE (2008) Note: using ecological theory to advance butterfly conservation. Isr J Ecol Evol 54:63–68

Shaffer ML (1981) Minimum population sizes for species conservation. Bioscience 31(2):131–134

Skinner CN (1995) Change in spatial characteristics of forest openings in the Klamath Mountains of northwestern California, USA. Landsc Ecol 10:219–228

St. Hilaire J, Bernatowicz J, Moore W (2007) U.S. Forest Service and Washington Department of Fish and Wildlife Mardon skipper (Polities mardon) Report. Naches Ranger District, Okanogan-Wenatchee National Forest, Spring and summer 2007. Unpublished agency report to USDA Forest Service and Washington Department of Fish and Wildlife, Naches

St. Hilaire J, Bernatowicz J, Moore W (2008) U.S. Forest Service and Washington Department of Fish and Wildlife Mardon skipper (Polities mardon mardon) Report. Naches Ranger District, Okanogan-Wenatchee National Forest, Spring and summer 2008, p 35. Unpublished agency report to USDA Forest Service and Washington Department of Fish and Wildlife, Naches

St. Hilaire J, Bernatowicz J, Moore W (2009) US Forest Service and Washington Department of Fish and Wildlife Mardon skipper (Polities mardon mardon). Report Naches Ranger District, Okanogan-Wenatchee National Forest, Spring and summer 2009. Unpublished agency report to USDA Forest Service and Washington Department of Fish and Wildlife, Naches

St. Hilaire J, Bernatowicz J, Moore W (2010) U.S. Forest Service and Washington Department of Fish and Wildlife Mardon skipper (Polities mardon mardon) Report. Naches Ranger District, Okanogan-Wenatchee National Forest, Spring and summer 2010, p 35. Unpublished agency report to USDA Forest Service and Washington Department of Fish and Wildlife, Naches

Stoner KJL, Joern A (2004) Landscape vs. local habitat scale influences to insect communities from tallgrass prairie remnants. Ecol Appl 14:1306–1320

Storm L, Shebitz D (2006) Evaluating the purpose, extent, and ecological restoration applications of indigenous burning practices in southwestern. Ecol Restor 24:257

Stoutjesdijk P, Barkman JJ (1992) Microclimate, vegetation and fauna. Opulus Press AB. Retrieved 2 Jan 2013, from http://www.cabdirect.org/abstracts/19930767963.html

The Urban Wildlands Group (2002) INCA: Insect Count Analyzer. Retrieved from http://www.urbanwildlands.org/INCA/

Thomas CD, Jones TM (1993) Partial recovery of a skipper butterfly (Hesperia comma) from population refuges: lessons for conservation in a fragmented landscape. J Anim Ecol 62:472–481

Thomas CD, Bodsworth EJ, Wilson RJ et al (2001) Ecological and evolutionary processes at expanding range margins. Nature 411:577–581

Thomas L, Buckland ST, Rexstad EA, Laake JL, Strindberg S, Hedley SL, Bishop JRB, Marques TA, Burnham KP (2009) Distance software: design and analysis of distance sampling surveys for estimating population size. J Appl Ecol 47:5–14

Thomas JA, Simcox DJ, Hovestadt T (2011) Evidence based conservation of butterflies. J Insect Conserv 15:241–258

Trunkle T, Fay P (1991) Transportation of spotted knapweed seeds by vehicles. Proceedings: Montana Weed Control Association, pp 14–16. Bozeman, MT

USFWS (2012) Endangered and threatened wildlife and plants; 12-month finding on a petition to list the mardon skipper as threatened or endangered; proposed rule. Fed Regist 77:54332–54352

Vaughan M, Black SH (2004) Report on Xerces Society Search and Monitoring for Polites mardon in Southern Oregon. The Xerces Society, Portland, June 22, 2004

Vaughan M, Black SH (2005) Report: Xerces Society Surveys for Polites mardon in the Rogue River National Forest. The Xerces Society, Portland, Summer 2005

Wagner D, Miller JC (1995) Must butterflies die for the gypsy moth's sins. Am Butterflies 3:19–23

Warren MS (1993) A review of butterfly conservation in central southern Britain: II. Site management and habitat selection of key species. Biol Conserv 64:37–49

Warren AD (2005) Butterflies of Oregon: their taxonomy, distribution, and biology. Lepidoptera of North America 6. Contributions of the CP Gillette Museum of Arthropod Diversity. Fort Collins, Colorado: Colorado State University, p 408

Whaley WH, Anhold J, Schaalje GB (1998) Canyon drift and dispersion of Bacillus thuringiensis and its effects on select nontarget lepidopterans in Utah. Environ Entomol 27:539–548

Widenmaie KJ, Strong WL (2010) Tree and forest encroachment into fescue grasslands on the Cypress Hills plateau, southeast Alberta, Canada. For Ecol Manag 259:1870–1879

Zald HSJ, Spies TA, Huso M, Gatziolis D (2012) Climatic, landform, microtopographic, and overstory canopy controls of tree invasion in a subalpine meadow landscape, Oregon Cascades, USA. Landsc Ecol 27(8):1–16

Habitat Restoration as a Recovery Tool for a Disturbance-Dependent Butterfly, The Endangered St. Francis' Satyr

Heather Cayton, Nick M. Haddad, Brian Ball, Erica Henry, and Erik Aschehoug

Abstract Habitat restoration is a key management strategy for species that experience loss, fragmentation or degradation of their habitat. The St. Francis' satyr is a federally endangered butterfly found only on Ft. Bragg Army Installation in North Carolina that depends on frequent disturbance to maintain its preferred habitat, ephemeral wetlands. Over the course of 4 years, we restored and maintained critical habitat for the St. Francis' satyr through a combination of hardwood removal and inundation via artificial dams to mimic natural beaver and fire disturbance. Here we present our insights into the challenges of creating and maintaining restored habitat for a species dependent on frequent disturbance, the results of experimental testing of demographic success, and the complexities in carrying out a captive rearing program. We discuss how species that are dependent on disturbance created habitat pose a unique challenge in creating and implementing a long-term plan to maintain a dynamic landscape.

Introduction

The destruction and degradation of native habitat caused by human activities represents the single greatest threat to biodiversity world-wide (Wilcove et al. 1998). To counter the loss of habitat, practitioners are increasingly turning to restoration as a key management strategy for the recovery of imperiled species. Restoration can be a strong tool for creating high quality habitat for a target species, while at the same time improving overall ecosystem health. However, habitat restoration actions have resulted in few successes (Suding 2011). In some cases, a lack of success is because there is no post-restoration monitoring program to assess recovery

H. Cayton (✉) • N.M. Haddad • E. Henry • E. Aschehoug
Department of Biological Sciences, North Carolina State University, Raleigh, NC, USA
e-mail: heather_cayton@ncsu.edu; nick_haddad@ncsu.edu; ehenry@ncsu.edu; erik_aschehoug@ncsu.edu

B. Ball
Endangered Species Branch, Directorate of Public Works, Fort Bragg, NC, USA
e-mail: brian.s.ball4.civ@mail.mil

© Springer Science+Business Media B.V. 2015

J.C. Daniels (ed.), *Butterfly Conservation in North America*,
DOI 10.1007/978-94-017-9852-5_8

(Bernhardt et al. 2007). In other cases, poorly defined restoration goals make it impossible to evaluate success (Zedler 2007). Restoration actions can also fail (Zedler et al. 2003; Holl and Hayes 2006), or even worse, create habitat that reduces rather than restores biodiversity (Breininger and Carter 2003; Robertson et al. 2013). Here, we present challenges and opportunities for the recovery of rare butterflies focused on restoration experiments.

A commonality shared by many rare butterfly species is the dependence on habitat created by disturbance. Restoration actions can be complicated when disturbance is required for the creation and maintenance of the desired habitat, as is necessary for many butterflies in need of protection. For example, the Schaus' swallowtail (*Heraclides aristodemus ponceanus*) is a federally endangered butterfly endemic to Florida that is subject to frequent tropical storms and hurricanes (Emmel 1995); although hurricanes may initially harm butterflies, they actually benefit from plant growth stimulated at the edge of hardwood hammocks. Fender's blue butterfly (*Icaricia icarioides fenderi*), a federally endangered taxon endemic to Oregon, lives in native grasslands where frequent fire is critical to stimulate growth of its host plant (Schultz and Crone 1998). What becomes immediately evident in any of these cases is that the same process that is required for butterfly persistence has the short term effect of reducing butterfly populations where the disturbance occurs.

In general, restoration programs focus on the creation of habitat with a single management action resulting in long-term, high quality habitat. Dynamic, disturbance dependent habitats present a challenge for managers because they require continued intervention in order to maintain high quality habitat. The integration of disturbance into restoration plans creates a unique challenge for managers, who must balance creating and maintaining habitat sufficient for recovery while also allowing for landscape scale processes, such as plant community succession, to continue.

Here we present our efforts to restore habitat and assess recovery success for the disturbance-dependent St. Francis' satyr (*Neonympha mitchellii francisci*), a federally endangered butterfly restricted entirely to Fort Bragg Military Installation in central North Carolina. Its primary habitat is herbaceous wetlands dominated by sedges (e.g. *Carex* spp.) along small riparian corridors. In North Carolina, this habitat is created when beavers abandon ponds, or by fires that reverse the rapid transition from open wetlands to closed canopy riparian forest (Kuefler et al. 2008; Bartel et al. 2010). However, the extirpation of most beavers from the region by humans a century ago, the suppression of fire, and increased human activity in the form of agriculture, development, roads and foot traffic, has eliminated or isolated much of the historic habitat. Because the habitat is naturally short-lived in space and time across the landscape, the St. Francis' satyr exhibits classic metapopulation structure, with small subpopulations that are distributed patchily within the landscape to form a single large population. The largest known subpopulations of St. Francis' satyrs are in artillery ranges on Fort Bragg, where a lack of roads, frequent small fires, and beavers have allowed for the persistence of high quality habitat. Little is known about the size of these subpopulations because highly restricted access to the sites precludes rigorous data collection. Outside of these areas, the number of subpopulations has been reduced from 5 to 2 in the last 15 years. In total, the remaining St. Francis' satyr habitat is thought to comprise 2–5 ha with an overall population size in the low thousands.

In order to maintain and increase the subpopulations of butterflies outside of the artillery target zones, we initiated a long-term restoration project in 2011 near remaining or historic St. Francis' satyr sub-populations. Through a combination of hardwood removal and inundation via artificial dams, we attempted to mimic the natural process of herbaceous wetland creation using a factorial experiment framework. Within our experimentally restored sites, we were restricted in the ways we could assess the response of St. Francis' satyrs to restoration because of its endangered status. Instead, we conducted detailed experiments on a closely related surrogate species to estimate butterfly population growth rates in different restored habitat types as a way to assess successful recovery. We also released captive reared adult butterflies into restored sites, and conducted transect surveys to track natural colonization of our restored sites over time. In this chapter, we present our insights into the challenges of creating and maintaining restored habitat for a species dependent on frequent disturbance, the results of experimental testing of demographic success, and the complexities in carrying out a captive rearing program.

Natural History

The St. Francis satyr is a small, brown butterfly that is a subspecies of *Neonympha mitchellii* (Fig. 1). The St. Francis' satyr is bivoltine, with adults emerging in late May through early June, and in late July through early August (Fig. 2). Extensive searching has determined that St. Francis' satyrs occur only on Fort Bragg military base, located in central North Carolina. Ft. Bragg is in the Sandhills region, and supports mostly longleaf pine forest with bottomland hardwood forests along stream floodplains. These streams dissect much of the terrain, but drainages are often interrupted by dirt roads used as fire breaks spaced every $\approx$ 200 m. Historically, fire played a large role in suppressing woody undergrowth along stream corridors and maintaining open herbaceous meadows. However, fire suppression is now common throughout the region. Current management plans at Fort Bragg include burning the pine understory approximately every 3 years, and largely exclude fire from riparian zones.

This stream network is also often modified by beavers, which dam portions of a creek to create flooded ponds that usually kill most standing hardwoods. Once these dams are abandoned and flood waters subside, the habitat is ideal for supporting wetland plants such as sedges in the *Carex* family. This includes *C. mitchelliana*, which is thought to be the main host plant for St. Francis' satyr larvae based on its successful use in captive rearing and its ubiquity in sites where St. Francis' satyrs are found. Other potential host plants that could support St. Francis' satyr larvae include *C. lurida* and *C. atlantica*, which larvae will eat in captivity, and *C. turgenscens*, which is widespread throughout the wetlands in artillery ranges.

Fig. 1 St. Francis' satyr (*Neonympha mitchellii francisci*), female, Fort Bragg, North Carolina (Photo by Melissa McGaw, North Carolina Wildlife Resources Commission)

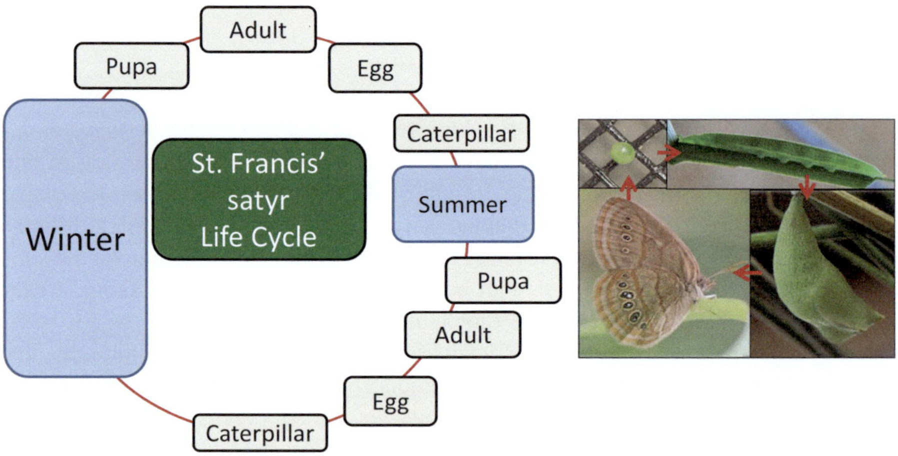

Fig. 2 Life cycle of the St. Francis' satyr (*Neonympha mitchellii francisci*)

Habitat Restoration

Our short-term goals upon starting restoration were threefold: (1) to document how restoration work impacted the plant communities in the wetlands, in particular sedge species; (2) to test whether hardwood removal, inundation, or a combination of both would best support St. Francis' satyr population growth; and (3) to establish populations of adult St. Francis' satyrs through the release of captive-reared individuals.

Long term data will better determine if restoration sites support growing or declining subpopulations, and how this may affect the fate of the overall population.

We focused restoration on two separate creeks outside the artillery ranges where St. Francis' satyrs have been found historically. Both creek systems are subject to woody encroachment, dry soils, and human activity that can quickly reduce the abundance of sedges. Historically, these two creeks supported two of the largest St. Francis' satyr subpopulations outside of artillery ranges, although numbers were low and in decline due to drought and woody encroachment when restoration began in 2011. As of 2014, no butterflies had been detected outside of restoration areas at either creek, indicating a need for urgent management.

A crucial aspect of restoration experiments is that they represent management actions that are practical. To insure this, we worked with local land managers to develop restoration treatments and methods that could easily be incorporated into future management actions at our study site. We restored wetlands using two techniques: manual removal of hardwoods and inundation. Both of these techniques represent a difficult trade-off for restoring habitat, as both are the result of natural disturbance and last for only a short time. In the case of the St. Francis' satyr, there is a relatively short period of time where habitat conditions are optimal. After disturbance, grassy wetlands quickly transition to hardwood forest, eventually reducing habitat quality for the butterfly. The inconsistent and random nature of disturbance in our system means that within a few seasons, the habitat must be re-evaluated to determine if disturbance is needed again. In light of these challenges, we planned our restoration activities with the knowledge that the effects of restoration were likely to be short-lived.

To create restoration sites, in the spring of 2011 we established four restoration areas, each containing four 30 m $\times$ 30 m experimental plots. In each area, plots were randomly assigned to one of four treatments in a factorial design: (1) manual removal of most trees; (2) installation of temporary dams; (3) both tree removal and installation of dams; and (4) no manipulation. Tree removal was intended to increase light availability within wetlands to encourage herbaceous plant growth. We intentionally left $\approx$ 10 % of the trees in order to mimic the canopy structure found in beaver-created wetlands in our area. We maintained tree removal treatments by cutting back the re-sprout of trees and shrubs in 2012 and 2013 to suppress hardwood canopy regeneration. Damming was intended to increase soil moisture and the amount of standing water, a key requirement for sedges. In dammed plots, we installed 0.5 m high and 1 m wide water-filled temporary coffer dams (Aquadam Inc., Scotia, CA, USA) across the length of the downstream edge of the plot, which inundated $\approx$ 10 m of the downstream portion of the plot. Within each plot we established vegetation subplots that were surveyed for percent cover of all sedge species and the major vegetation types (e.g. grasses, forbs, and shrubs) prior to restoration in Spring 2011, as well as in Fall 2011, Spring 2012, and Spring 2013.

Restoration of wetland habitat via the removal of trees had strong effects on the plant community. Tree removal had a significant positive effect on sedge populations, with the average percent cover of *Carex mitchelliana*, an important host plant for St. Francis' satyr, more than doubling (from 3 to 7 %) and the percent cover of all sedge species combined more than quadrupling (from 7 to 31 %) from 2011 to 2013. The percent cover of all understory vegetation also increased significantly in tree removal plots, with an increase from 34 to 73 %. In contrast, damming did not have a significant positive effect on the percent cover of *C. mitchelliana*, all sedges, or all understory vegetation.

By experimentally testing each restoration treatment independently, we were able to identify tree removal as the most important factor determining host plant size and abundance in our system. The increases in sedge and understory cover are likely due to increases in light availability. It is important to note that increases in host plants do not guarantee butterfly population recovery success, but do create a favorable landscape for individuals to become established.

The disturbance-dependent nature of these herbaceous wetlands means that our results, while positive, do not necessarily represent a long-term, stable change to the plant community. Without continued intervention, restoration sites are likely to revert back to original conditions, with woody encroachment reducing the herbaceous understory. There is a challenging balance between constant maintenance of host plant dominated communities, and allowing natural regeneration of trees to occur. Ultimately, the amount and frequency of management actions can only be determined by understanding the population dynamics of St. Francis' satyrs over time.

Assessing Restoration Success

Deciding how to measure the success of restoration is as critical as determining what the restoration actions will be. While presence/absence counts of a target species are easy to conduct with minimal time and effort, they give little indication of whether the dynamics within the restored area are favorable to long-term success. Even population counts, which superficially give a measure of success by showing quantitative changes in population size, provide little information on whether populations are growing due to recruitment via immigration or reproduction within the site. Thus, it is vital to measure demography in addition to population size. Demographic rates provide information on whether a population is growing (i.e. the number of births exceeds the number of deaths) without the complication of considering immigration from outside areas. The ultimate goal of restoration is to create stable, increasing populations; therefore, success can only be determined by explicitly measuring population growth rates independent of immigration and dispersal.

To understand more about the population growth rates of St. Francis' satyrs in our restoration plots, we experimentally estimated the survival rates of eggs and juveniles of the closely related surrogate species Appalachian Brown butterfly (*Satyrodes appalachia*). The Appalachian Brown is a locally rare, bivoltine butterfly that prefers *Carex* species as a host plant, and is found in the same herbaceous wetlands as St. Francis' satyrs (Hudgens et al. 2012) at our study site. We chose to use a surrogate species because of the limited population size of St. Francis' satyr, and because of the tremendous risks restoration experiments conducted with an endangered species pose to individuals and the population as a whole. However, the use of a surrogate species, no matter how similar in behavior and biology, runs the risk of not accurately representing how the imperiled species will respond to restoration. Therefore, experiments using a surrogate species should be, when possible, coupled with additional measures of success that relate directly to the target species such as population counts via detailed surveys.

Determining population growth rates, whether using a surrogate or not, is complicated by the cascading effect restoration can have on the environment. Changes can occur in both biotic factors, such as host plant abundance or quality, and abiotic factors, such as temperature and light conditions. These changes can directly impact the target species by creating high quality habitat for development and breeding; however, these changes can also have unintended indirect effects, for example by improving habitat conditions for predators. For St. Francis' satyrs and Appalachian Browns, an open canopy provides additional light for *Carex* growth, but also makes it easier for known aerial predators such as dragonflies to prey on adult butterflies, and for ants and spiders, which prey on eggs and small larvae, to move and hunt. Therefore, it is important to determine how restoration activities both directly and indirectly impact different life stages of the target species, and strike a balance between habitat that supports butterflies but does not encourage predators to proliferate.

To estimate the direct and indirect effects of restoration treatments on *S. appalachia* demography, we estimated egg survival and larval to adult survival independently. To estimate egg survival, we placed greenhouse raised, potted *C. mitchelliana* plants with a known number of eggs into our established experimental restoration plots and counted the number of eggs that remained after 48 h. We first caged wild-caught *S. appalachia* females on potted sedge plants in a greenhouse at Fort Bragg, and counted the number of eggs laid. Each restoration plot received one pair of potted sedges, one randomly assigned to a predator exclusion treatment and a second that was accessible to all predators. Plants in the predator exclusion treatment were enclosed in a cage constructed from a 20 L fine mesh fabric paint strainer and bent wire. A 15-cm band of Tanglefoot® insect barrier was applied to pots to exclude crawling predators. Paired plants were placed in the center of each restoration plot on dry ground within 5 m of each other. All plants were removed from the field after 48 h and we counted the number of eggs remaining.

To estimate survival over the larval and pupal stages, we deployed a known number of larvae into experimental arenas created in each plot and counted the number of butterflies that emerged. We constructed larval arenas by removing the ends of a standard 55 gal polyethylene food grade drum and cutting the remainder into thirds, resulting in an approximately 57 cm diameter × 35 cm tall ring. At six randomly chosen locations in each plot, rings were buried approximately 10 cm into the ground surrounding naturally occurring, mature *C. mitchelliana*. In one block with low sedge abundance, field cages were placed around greenhouse raised and transplanted *C. mitchelliana* plants. We randomly assigned each arena to a predator-free or predator-accessible treatment. Predator-free arenas were enclosed with tulle netting that was secured to the outside of the rings using nylon straps and tied at the top of a vertical support structure to hold the fabric off the plants. All potential predators were manually removed from predator-free arenas prior to addition of larvae. Arenas accessible to predators remained open during larval development, but once pupae were observed, arenas were enclosed with tulle netting to facilitate capture of emerging adults. Five larvae varying in age from first to third instars (most commonly second instars) were placed into each arena in a block on the same day. Larvae were left to develop in the field. After the observation of pupae, arenas were checked daily for newly emerged adults until no butterflies were found for five consecutive days.

Despite the fact that our restoration treatments had strong, differential effects on our wetland systems, such as increases in sunlight, water, and host plants, we found that these differences did not affect the survival of eggs, larvae or pupae when protected from predation. This suggests that common mortality factors in early stage development like egg desiccation and plant quality are likely not important drivers of population growth rates of *S. appalachia.* It also suggests that the location of a host plant is not as important as the presence of the host plant. *S. appalachia* appear to be robust to a wide range of water and temperature regimes as they develop, and potentially can be successful as long as they have access to their host plant without excessive predation.

Another important finding from our experiments is that restoration treatments had different effects on the levels of predation experienced by eggs and juveniles. When eggs were exposed to predation, survival in the three types of restored plots was about one-third the survival in control plots, meaning our manipulation of tree cover and standing water appears to have substantially increased predation of butterfly eggs. In contrast, when juveniles were exposed to predation, survival rates were different depending on the type of restoration. Survival in dam only plots was twice that of both control plots and plots where both trees were removed and dams installed, and five times that of tree removal only plots. The differences in predation among restoration treatment types appear to be driven by changes in access to juveniles by ground predators. The installation of dams increased the amount of standing water within our plots and likely reduced the ability of ground predators, such as ants and spiders, to gain access to juveniles.

Interestingly, these predation effects appeared to outweigh any potential benefit butterfly populations might gain from the restoration of plant communities. Tree removal treatments tripled potential host plant cover, but also led to the highest rates of predation for both eggs and juveniles. The addition of dams decreased predation of juveniles, but did not increase the abundance of host plants. Most importantly, we found that no single restoration treatment both increased host plant abundance and egg or juvenile survival. These results highlight the need for experiments that carefully measure the recovery potential of different restoration methods across all life stages of butterflies. In our case, because of the conflicting effects we found, further work is needed to fully understand how our restoration treatments impact population growth rates. Only when this is complete can we expect to guide management decisions for restoration.

Population Expansion: Natural Colonization

Even if restoration actions result in the creation of habitat that supports growing populations of butterflies, the establishment of new populations first requires successful colonization. Ideally, individuals are able to reach restored habitat via natural dispersal because this allows the overall population to both grow at a sustainable rate, and ensures that individuals can move freely between subpopulations, which is important for maintaining genetic diversity. Where restored habitat is in close proximity to existing populations, or can be connected via corridors, the likelihood of natural colonization may be high. Detections of natural colonization are a critical first measure of success, and support the need for close monitoring to determine where and when individuals arrive.

In May 2012, the first flight period after restoration activities were completed, we detected the first St. Francis' satyr adult in a restoration plot. This individual was undoubtedly a wild individual, because St. Francis' satyr were not present in restoration sites prior to restoration and no captive reared individuals were ever released into this restoration site. However, there is no way to distinguish whether it emerged in the restoration plots as a result of a prior colonization event or dispersed there as an adult from the historic colony that is located less than 250 m away. This natural colonization event was followed by several other detections of individuals during that flight period and in subsequent flight periods. The continued detection of individuals over time confirms that natural colonization by St. Francis' satyrs is possible between our restoration sites and existing subpopulations.

Following these colonization events, we were able to use daily transect surveys to assess the overall population trend across multiple years (Fig. 3). While populations in non-restored areas remained relatively stable, populations in restored areas showed an increase each year. This information was a valuable first step in determining whether restoration is successful in sustaining a growing population,

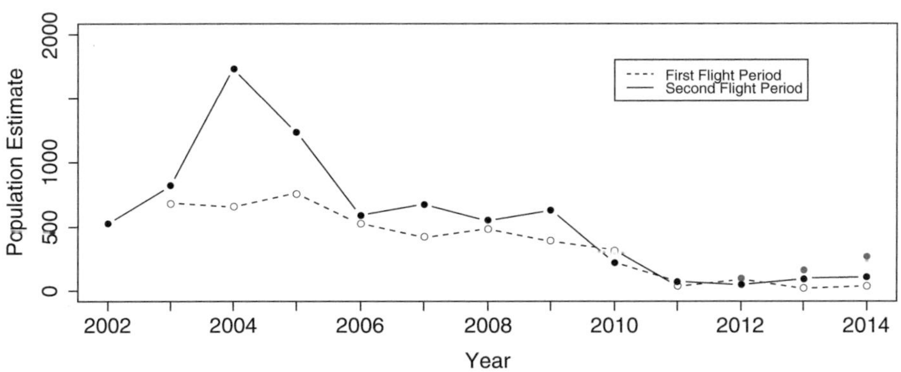

Fig. 3 Population size of St. Francis' satyrs outside of artillery ranges on Fort Bragg from 2002 to 2014. *Red points* in years 2012–2014 represent the overall population size when individuals from restoration sites are included

and represents more than we could have determined with simple presence/absence surveys. When combined with demographic data from the surrogate species, it allows us to gain a much broader view of overall population dynamics. We are now in a position to not only measure success in a quantitative way, but also have a better sense of what future population trends will look like.

Population Expansion: Captive-Reared Releases

Although natural colonization is ideal, in many cases it is unlikely or even impossible. Impediments such as low population size, limited dispersal, human development, or poor surrounding habitat conditions may prevent individuals from moving into restored areas on their own. In such cases, artificially colonizing sites through the release of captive-reared individuals may be the best strategy to populate a restored area. For the St. Francis' satyr, restoration on one creek allowed for natural colonization. However, the restoration area on the second creek was so isolated from other existing populations that natural colonization was not possible. We considered the release of captive-reared St. Francis' satyrs to be critical to the success of restoration in this scenario, so we began a captive-rearing program to create new subpopulations and augment the existing overall population.

In 2012, we expanded our captive-rearing program into a large greenhouse on Fort Bragg in order to control temperature, exclude predators, and ensure a constant population of host plants. We maintain the greenhouse at approximately ambient temperature year-round, and keep a continuous stock of mostly *C. mitchelliana* available for larval growth and adult breeding. To generate a captive stock, we collect 3–5 females from artillery ranges during both flight periods, when possible. Females were kept for up to 72 h in small oviposition chambers in order to collect eggs before being released at one restoration site. All eggs were placed in a small dish at the base of a *C. mitchelliana* plant. We allowed newly hatched caterpillars to crawl onto the plant unassisted. This process was closely monitored; we manually assisted caterpillars that were unable to find a leaf on their own. Once on the plant, caterpillars developed without further assistance. Captive rearing enclosures were monitored daily for the presence of predators and to ensure all *C. mitchelliana* plants were healthy. Because we carefully controlled the greenhouse temperature to match our site conditions, captive-reared adults emerged at approximately the same time as the wild population. We marked all individuals with a unique ID and released them at a restoration site within 24 h of emergence.

We successfully raised St. Francis' satyrs from egg to adult stage for every flight period where we were able to collect wild eggs (Table 1). Wild females collected from artillery ranges laid an average of 16 eggs per female, although not all females laid eggs, indicating that we captured some females prior to mating. Although the success rate for eggs transitioning to caterpillars was variable, our success rate for obtaining viable adults once individuals survived to the caterpillar stage averaged 56 % when years when no eggs were laid are excluded.

Table 1 Results of captive-rearing of St. Francis' satyrs from 2011 to 2014

Year	Flight period	Captured females	Eggs	Caterpillars	Adults	Survival rate
2011	1	3	60	10	9	0.90
	2	5	142	86	32	0.37
2012	1	4	65	59	27	0.46
	2	0	0	0	0	0.00
2013	1	4	102	76	47	0.62
	2	0	0	0	0	0.00
2014	1	5	122	62	28	0.45
	2	4	88[a]	70[a]	NA	NA

Eggs were collected from wild females captured in artillery impact areas. Years with no eggs were a result of no trips taken into artillery impact areas during the flight period
[a]Includes eggs and caterpillars from one captive female who successfully mated

Table 2 Population estimates based on Pollard-Yates indices of four restoration plots from 2012 to 2014

Year	Flight	Captive adults released	Wild females released	P-Y index (unmarked)	P-Y index (unmarked and marked)	Population estimate
2012	1	0	0	0	0	NA
2012	2	27	0	0	1.3	11
2013	1	0	4	0.45	0.45	NA
2013	2	47	0	5.42	9.07	106
2014	1	0	5	3.00	3.00	32
2014	2	28	4	12.25	14.65	175

Some flight periods had too few individuals to estimate population size

We released captive-reared individuals into the same site each flight period beginning in the second flight period of 2012 (Table 2) and detected unmarked individuals there during each subsequent flight period. These unmarked individuals represent wild individuals that are almost certainly the result of successful breeding within the restoration site because the only existing subpopulations are too far away to have such a large and immediate effect via dispersal and colonization. We detected an increase in population size during each successive year through the use of transect surveys. Our most recent estimate is of almost 200 individuals across our restoration site, which is approximately 1/2 ha in size.

In the 3 years since restoration began, we have noticed a distinct increase in the overall St. Francis' satyr population, owing mostly to individuals seen in restored sites (Fig. 3). This increase follows several years of decline in historic sites, where populations still continue to decline due to dry conditions and woody encroachment. Without the additional restoration undertaken in 2011, it is possible that St. Francis' satyrs may have gone extinct at all sites outside of the artillery ranges. Our current positive trend shows great promise for the successful restoration of St. Francis' satyrs, although long term data will be needed to determine how long habitat persists after intervention, and whether the overall population will continue to grow naturally.

Summary and Key Points

- Restoration can be an important step in managing for rare and endangered species that encounter habitat loss or fragmentation, especially when conducted within a factorial experimental framework. Without a clear structure for determining which specific restoration activities are effective, it is difficult to fully understand which actions contribute to population growth.
- Species that are dependent on disturbance created habitat pose a unique challenge in creating a long-term plan to maintain a dynamic landscape.
- Disturbance-dependent species such as the St. Francis' satyr can be managed with multiple restoration techniques such as inundation and hardwood removal, but long term success requires monitoring to determine the continued effectiveness of the restoration actions.
- Captive-rearing can be a highly effective method of augmenting existing populations and artificially colonizing restored sites that may otherwise see little or no dispersal from already established populations.
- Success is best determined by closely measuring the population growth rates of butterflies in restored sites. However, other measures, such as adult survivorship, breeding capacity, or vegetation growth may serve as a suitable proxy. Importantly, goals should be clearly outlined at the start of a restoration program so that the appropriate measures of success can be implemented.
- Restoration requires several years of monitoring to determine its effectiveness, especially with rare or endangered butterflies, where a short time span of only one flight period or one season is unlikely to accurately assess population trends. Continuous monitoring over several seasonal cycles is the only way to determine population trends, which are the best indication of whether restoration has been successful.

References

Bartel RA, Haddad NM, Wright JP (2010) Ecosystem engineers maintain a rare species of butterfly and increase plant diversity. Oikos 119:883–890

Bernhardt ES, Sudduth CB, Palmer MA et al (2007) Restoring rivers one reach at a time: results from a survey of U.S. river restoration practitioners. Restor Ecol 15:482–493

Breininger DR, Carter GM (2003) Territory quality transitions and source–sink dynamics in a Florida scrub-jay population. Ecol Appl 13:516–529

Emmel TC (1995) Habitat requirements and status of the endemic Schaus swallowtail in the Florida Keys. Florida Game and Fresh Water Fish Commission, Nongame Wildlife Program, Tallahassee

Holl KD, Hayes GF (2006) Challenges to introducing and managing disturbance regimes for Holocarpha macradenia, an endangered annual grassland forb. Conserv Biol 20:1121–1131

Hudgens BR, Morris WF, Haddad NM et al (2012) How complex do models need to be to predict dispersal of threatened species through matrix habitats? Ecol Appl 22:1701–1710

Kuefler D, Haddad NM, Hall S et al (2008) Distribution, population structure and habitat use of the endangered Saint Francis Satyr butterfly, Neonympha mitchellii francisci. Am Midl Nat 159:298–320

Robertson BA, Rehage JS, Sih A (2013) Ecological novelty and the emergence of evolutionary traps. Trends Ecol Evol 28:552–560

Schultz CB, Crone EE (1998) Burning prairie to restore butterfly habitat: a modeling approach to management tradeoffs for the Fender's blue. Restor Ecol 6:244–252

Suding KN (2011) Toward an era of restoration in ecology: successes, failures, and opportunities ahead. Annu Rev Ecol Evol Syst 42:465–487

Wilcove DS, Rothstein D, Dubow J, Phillips A, Losos E (1998) Quantifying threats to imperiled species in the United States. Bioscience 48:607–615

Zedler JB (2007) Success: an unclear, subjective descriptor of restoration outcomes. Ecol Restor 25:162–168

Zedler JB, Morzaria-Luna H, Ward K (2003) The challenge of restoring vegetation on tidal, hypersaline substrates. Plant and Soil 253:259–273

Butterflies Are Not Grizzly Bears: Lepidoptera Conservation in Practice

Travis Longcore and Kendall H. Osborne

Abstract Many of the well-known and intuitive practices from conservation of large mammals do not apply to butterflies. The different body size, lifespan, reproductive potential, and habitat associations of butterflies produce a different series of guidelines for conservation that may conflict with commonly held attitudes and approaches from conservation of larger species. Using a series of case studies from the federally listed butterflies in California, we present a series of lessons for butterfly conservation that emphasizes the different approaches that butterfly conservation requires. These include: that habitats found surrounded by cities are important; that connectivity between habitats may not require a corridor or can be provided by human intervention; that disturbance, including removal of native plants, may be essential to maintain butterfly food plants; that a remote risk of incidental take of early stages of butterflies is often worth taking in the interest of habitat management; that current distributions of taxa may not be their best habitat historically; and that microclimatic and microtopographic variation is important at scales humans rarely consider.

Introduction

Perhaps because of the conspicuous nature and broad appeal of butterflies to the public, their plight as threatened by human activities has resulted in many species receiving formal legal protection. They capture the imagination, even if the only species that the general public might recognize is a Monarch (*Danaus plexippus*). Interest is widespread, and the release of an endangered butterfly to a newly restored habitat generates considerable media attention.

T. Longcore (✉)
School of Architecture and Spatial Sciences Institute, University of Southern California, Los Angeles, CA, USA

The Urban Wildlands Group, Los Angeles, CA, USA
e-mail: longcore@usc.edu

K.H. Osborne
Osborne Biological Consulting, Riverside, CA, USA
e-mail: euproserpinus@msn.com

© Springer Science+Business Media B.V. 2015

J.C. Daniels (ed.), *Butterfly Conservation in North America*,
DOI 10.1007/978-94-017-9852-5_9

The everyday work of butterfly conservation, and indeed of conservation planning for most terrestrial invertebrates, is taken up by local, state, and federal agency staff, nongovernmental conservation organizations (including those holding and managing land), professional consultants, the occasional academic researcher, and interested members of the public. Not every person assigned or volunteering to work on butterfly conservation has training in entomology, yet most probably have a general background in conservation biology and its principles. The most prominent principles of conservation biology, which are both taught at university and reasonably easily understood by the general public, do not always apply to butterflies or do so at a scale at which people are not used to thinking. That is, the prescriptions of conservation biology that apply to grizzly bears, a need for large blocks of undisturbed habitat with contiguous corridors and attention to careful enumeration of individuals (e.g., Wielgus 2002), do not apply to most butterfly species.

A significant part of population biology theory derives from the study of butterflies, with checkerspot species on two continents serving as focal species for the development of metapopulation theory (Hanski 1999; Ehrlich and Hanski 2004). Such work, and our combined practical experience and research over four decades on the conservation of butterflies in California and beyond, can help to re-scale butterfly conservation from the simple yet appealing lessons applicable to grizzly bears and other large, long-lived species to approaches more appropriate for small, short-lived species with potentially explosive population growth.

Many species of butterflies are similar to grizzly bears of centuries past in that they ranged widely across the North American continent. But unlike large mammals, butterflies are prone to local specialization to habitats, foodplants, and microclimatic conditions to an extent that large-bodied and long-lived species are not. As a result, these wide-ranging butterfly species have multitudes of local subspecies or ecotypes that are subject to and respond to local selection pressures. In California in particular, the topographic diversity, varied coastline, and many disjunct and diverse vegetation communities led to the evolution of many locally adapted subspecies on coastal bluffs and dunes, peninsulas, grasslands, and other unique habitats. Many of the "species" in danger of extinction today are these localized and genetically unique subspecies of more broadly ranging species.

The purpose of this chapter is to identify key ways in which rules of thumb or "common knowledge" approaches to conservation must either be adjusted or interpreted differently to manage and restore populations of endangered butterflies. The iconic examples of endangered conservation efforts that are known to the public are predominantly centered on large, long-lived species (often predators) with associated large home ranges, such as wolves, jaguars, grizzly bears, and California Condors. In the sections that follow, we propose six ways in which butterflies differ: (1) cities matter; (2) connectivity does not need to be contiguous; (3) disturbance, even from humans, can be essential for survival; (4) geographic distribution is more important than population numbers (with some caveats); (5) remnant habitat is not necessarily high quality habitat compared with historical distributions; and (6) a molehill can be a mountain, especially to a butterfly larva. We begin the chapter with species accounts for several federally listed butterfly species in California with which we have experience and that are then used to illustrate the lessons that follow.

A Primer of Listed California Butterflies

Fourteen species of butterflies found in California are federally listed under the Endangered Species Act. We review the life history and broad brushes of evolutionary context in which interpretation of their status and their prospects for recovery must be considered. They are divided into two groups, one for southern California and the other for the San Francisco Bay region, and we highlight species with which we have some experience.

Southern California

Southern California listed species range from a climatic relict (Laguna Mountains Skipper) to two coastal subspecies (El Segundo Blue and Palos Verdes Blue) to a previously widespread subspecies extirpated from the core of its range (Quino Checkerspot).

Laguna Mountains Skipper

Laguna Mountains Skipper (LMS; *Pyrgus ruralis lagunae*) is an isolated subspecies of the more widely distributed Rural Skipper (*Pyrgus ruralis*), and its endangered status may be better understood when examined in the context of the species as whole. Rural Skipper ranges through western North America from British Columbia to California. In central California, populations are found from the coast at Santa Cruz, clear up to high elevations in the Sierra Nevada. Throughout this range butterflies use several different species of *Potentilla* and *Horkelia* (Rosaceae). Larvae feed, form nests, and pupate within leaves of these food plants, such that larvae and pupae are typically situated several centimeters above the ground within the host foliage.

Twenty thousand years ago, the climate at what is now Los Angeles, California was similar to that of Santa Cruz—with madrone, redwoods, and Monterey pine (Axelrod 1978), and the rural skipper was no doubt more broadly distributed in the area, even down to the coast. In southern California, climatological changes, drying, and the upward elevational migration of wetter climates led to the acute isolation of these skippers on the mountain meadows of the Peninsular Ranges, where the butterflies evolved to specialize on their local host *Horkelia clevelandii*, and became the distinct subspecies that we know today. The butterfly was restricted to the Laguna Mountains (apparently extirpated) and Palomar Mountain (extant) of San Diego County, the two populations now so distant from one another that gene flow could not occur even if both areas were occupied.

For over a century, the lands have been subject to cattle grazing. The practice became especially intense during the twentieth century on private ranch lands and as part of the economic uses policy pursued by the Cleveland National Forest (United States Department of Agriculture 2005). Cattle feed close to the ground, and consistently crop the *Horkelia clevelandii* down to a height of 2 cm, so that LMS larvae and pupae would be subject to ingestion in the course of grazing events. Ingestion by cow likely represents a significant mortality factor new to the ecology of LMS, a factor that has been relentlessly pressuring the LMS populations for decades. Although the butterfly persists with limited distribution on Palomar Mountain, it has not been documented in the Laguna Mountains since 1999 where the meadows have been under heavy grazing pressure for decades.

LMS was federally listed in 1997 (U.S. Fish and Wildlife Service 1997). Many years of intense survey efforts continue to produce negative results for the butterfly on the Laguna Mountains. By 2002 the National Forest fenced several locations where *H. clevelandii* is abundant as part of their management responsibilities (Osborne 2003). Over the years, where cattle have been excluded, the *Horkelia* has made a dramatic comeback, with foliage of many plants extending 10–30 cm above the ground. Unfortunately, the habitat protection came too late because the skipper had already been extirpated. The recovered food plant stands in the Laguna Mountains appear entirely suitable for the butterfly and a reintroduction of the butterfly from the remaining population at Palomar Mountain is an obvious remedy. LMS eggs are easy to obtain from wild females and could be the basis of a reintroduction, but such an effort has not yet been permitted. Incremental progress has been made, however, with captive rearing of the species as a means to produce stock for reintroduction (Johnson et al. 2010; Longcore et al. 2014).

Palos Verdes Blue

The Silvery Blue (*Glaucopsyche lygdamus*), diversified into a dozen subspecies, ranges from arctic Alaska across Canada to Newfoundland and across the northern temperate zones with little variation, then south along the Appalachian chain, and throughout the western continent south to northern Baja California. Geographic variation is most prevalent in the West. They feed on various Fabaceae including *Lupinus*, *Acmispon* [=*Lotus*], *Astragalus*, and *Vicia*. Larvae produce honeydew and have a facultative relationship with tending ants.

One subspecies, the Xerces Blue (*Glaucopsyche lygdamus xerces*) used to fly on coastal dune habitats of the upper San Francisco Peninsula, where its larvae fed on *Acmispon glaber* [=*Lotus scoparius*]. It was driven to extinction after 1943 in the wake of urban development. The Palos Verdes Blue (PVB; *Glaucopsyche lygdamus palosverdesensis*) is a peninsular subspecies in southern

California and, like the Xerces Blue, most of its habitat has been subsumed by urban development. Restricted to the Palos Verdes Peninsula, it was considered remarkable for its specialization on the locoweed, *Astragalus trichopodus lonchus*, as a larval foodplant. The PVB was federally listed as endangered in 1980, and in 1982 the City of Rancho Palos Verdes bulldozed the last known remaining population for recreational facilities, simultaneously ridding itself of a troublesome federally protected species within its jurisdiction. In 1994, the butterfly was discovered on the disturbed grounds of a military installation, the Defense Fuel Support Point, San Pedro (DFSP) at the base of the peninsula where it used *Acmispon glaber* as host in addition to the locoweed. PVB is closely related to southern blue (*Glaucopsyche lygdamus australis*), which is broadly distributed through wildland areas of southern California outside of the Palos Verdes Peninsula and also uses *Acmispon glaber* as host.

PVB at DFSP had persisted for decades in the context of high disturbance and abundance of its early-successional *Acmispon glaber* host. After its discovery in 1994, habitat protection, restoration, and captive breeding were implemented (Mattoni 1995; Mattoni et al. 1998; Mattoni and Longcore 2002). Numbers appeared to be stable and increased slightly through 2003, but with the maturation of coastal sage scrub habitats and associated successional decline of *Acmispon*, the measurable adult butterfly population has declined to zero in the most recent surveys (Fig. 1). Severe drought conditions in 2014 and 2015 may have depressed the flight, and a diminishing possibility remains that wild butterflies remain in multiple year diapause as pupae. A captive breeding program is ongoing and available for reintroduction.

From the captive colony, PVB has been introduced to three other sites on the peninsula outside of the DFSP, but these reintroductions have not yet resulted in a documented self-sustaining population. At Friendship Park, only a few mature *Astragalus trichopodus* are found, and virtually no *Acmispon glaber*, and the butterfly has not been able to become established. At an undisclosed coastal site with *Acmispon* (which soon phased out with vegetation succession), some adults were seen for a couple of years flowing release, but not since. At the Linden H. Chandler Preserve, where *Acmispon* and *Astragalus* have been planted with some success, multiple releases have led to small numbers of adults seen, but not yet documented beyond a second year. The DFSP remains the preeminent site for future conservation of the PVB.

El Segundo Blue

The Square-spotted Blue (*Euphilotes battoides*) ranges across western North America as many subspecific entities, each specialized on various relatively more locally distributed species of buckwheat (*Eriogonum*; Polygonaceae). Larvae feed on the blossoms, secrete honeydew, and thereby draw the attentions of tending ants, which in turn ward off many predators and parasitoids of the larvae (Mattoni 1992, 1993).

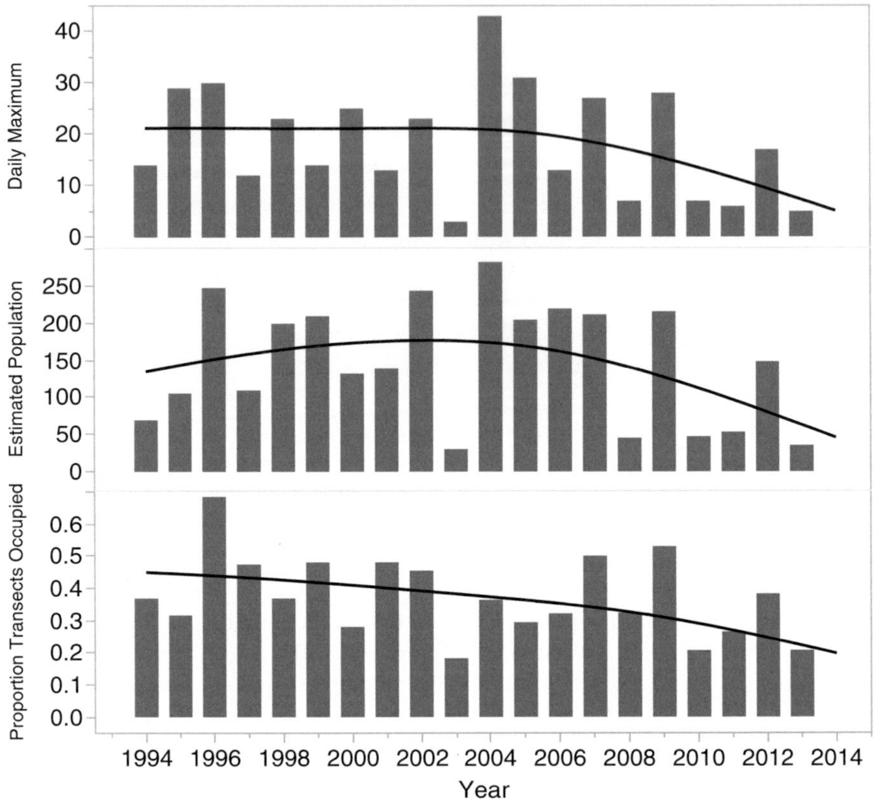

Fig. 1 Daily maximum, estimated population, and percent of transect segments occupied for Palos Verdes Blue at Defense Fuel Support Point, San Pedro, 1994–2014 (Data from Longcore and Osborne 2013). Smoothed curve fit is from kernel density estimation. Long-term decline in proportion of transect segments occupied has less annual deviation attributable to "good" and "bad" weather years

Although this ant/larva association, common across lycaenid butterflies worldwide, is not obligatory for *Euphilotes*, it obviously provides a selective advantage (hence the lycaenid butterflies maintain the larval structures to secrete honeydew). Seacliff buckwheat (*Eriogonum parvifolium*) is limited to the immediate coast of southern California, especially on beach and dune habitat and to a lesser extent on coastal cliffs; the *Euphilotes* species specialized on, and obligate to this host, is El Segundo Blue (ESB), a.k.a. Allyn's Blue (*E. battoides allyni*).

Beachfront properties are high-end real estate and were quickly developed to residential and commercial use as the City of Los Angeles and its satellite communities grew in the early twentieth century. In other areas of coastal dune not developed, intense recreational beach use, and invasion of exotic plants (especially *Carpobrotus edulis*; ice plant), displaced the native buckwheat from many undeveloped beach habitats. By the time the butterfly was federally listed in 1976 (U.S. Fish and Wildlife Service 1976) it was known only on a small lot next to oil refineries in the City of El Segundo. ESB was subsequently found to persist also on coastal dunes at the west end of the Los Angeles International Airport, and on tiny relictual enclaves on the Palos Verdes Peninsula farther south (Mattoni 1992; Herrmann 1999; Osborne 2001); and now a population of butterflies at Vandenberg Air force Base in coastal Santa Barbara County is also considered to be ESB (Pratt and Stouthamer 2008) although the mechanism by which such a disjunct distribution came to be established is not clear.

Mattoni et al. (2001) found the highest densities of ESB on the lee (protected) side of dune slopes on a scale of hundreds of meters, where the vegetation and butterflies were relatively protected from relentless marine onshore winds. Similarly, on a scale of meters, high densities of ESB were present in the protective lee of large shrubs and boulders at the crest of marine bluffs (Osborne 2001). Although these relatively protected backdune areas are rare, restoration of native vegetation and *E. parvifolium* on open beach dune and bluff within small plots on State and County public beaches, has led to the dramatic resurgence of ESB in large numbers (Schoch 2007). These results, which include natural recolonization of over 4 acres of beach bluff habitat in Torrance and Redondo Beach, recolonization of a remnant dune at the Ballona Wetlands, and colonization of restored coastal strand/foredune habitat at Dockweiler State Beach, provide hope that ESB can recover from the brink of extinction with more extensive habitat restoration and ongoing stewardship.

Quino Checkerspot

Edith's Checkerspot (*Euphydryas editha*) is a nymphalid butterfly with a broad range across western North America from British Columbia south to Baja California Norte with many subspecies described from various mountain ranges, elevations, and sectors of the Pacific coastline, including at least 18 subspecies from California alone (Garth and Tilden 1986; Emmel 1998). Local populations tend to be specialized on locally abundant hosts (White 1974; Scott 1986), consistently and broadly among the Scrophulariaceae and in the closely related families Orobanchaceae and Plantaginaceae *sensu lato* [now Scrophulariaceae]. Various subspecies have been described from coastal California, along what is essentially a gradational cline from dark butterflies (with heavy black scales) in the north (San Francisco Bay area) to mostly red-orange butterflies in the southern end of the range in Baja California. Two of these, the Bay Checkerspot Butterfly (*E. e. bayensis*) and the Quino Checkerspot Butterfly (QCB; *E. e. quino*) are listed under the Endangered Species Act.

Populations of the Bay Checkerspot are restricted to open grassland/forbland communities on a series of geologically isolated outcrops of serpentine and associated serpentine clay soils. Loss of too many serpentine hills to urban development imperiled this butterfly. Serpentine soil chemistry largely restricts many exotic grasses, trees, and shrubs, and allows the short-lived annual food plant, *Plantago erecta*, along with many native wildflowers, to grow in abundance in habitat that is reminiscent of alpine meadow. In southern California, where serpentine geology is absent, QCB colonies are primarily associated with low elevation (sea level to 3,000 ft) open grasslands, vernal pools, and sunny openings within chaparral, coastal sage scrub, and juniper woodlands. Colonies are found frequently associated with volcanic gabbroic, basaltic, or andesitic clay soils (Osborne unpub. obs.) and/or undisturbed soil surfaces that possess cryptogamic crusts (Osborne and Redak 2000; Mattoni et al. 1997). QCB was probably once the most abundant springtime butterfly in southern California when *Plantago erecta* would have been ubiquitous, growing unfettered by yet-to-be introduced exotic grasses and forbs (Mattoni et al. 1997). The introduction of these weeds by Spanish explorers and missionaries of the sixteenth century forever changed the floral ecology of California generally, and particularly reduced the extent and abundance of *Plantago erecta* by competitive exclusion (Proctor and Woodell 1975). As with the Bay Checkerspot, long after habitat restrictions imposed by these floristic conversions, QCB has been declining since the late 1960s primarily due to habitat loss from urban and agricultural expansion (Thorne 1970; Emmel and Emmel 1973; Orsak 1988). Large-scale habitat loss additionally disrupts metapopulation stability leading to the eventual collapse of the entire species on a regional scale. Fire and overgrazing (Orsak 1977, but see Orsak 1988) may explain additional decline.

Metapopulation dynamics (Ehrlich et al. 1980; Dobkin et al. 1987) are an important element of checkerspot population ecology. Local colonies of checkerspots exist within complexes of habitat patches with many habitat patches seeing temporary and often long-term absence of butterflies. Populations undergo irregular boom and bust cycles. Particular, large patches of high quality habitat serve as population sources from which adults disperse to recolonize outlying habitat patches. Habitat patches vary with respect to food plant abundance, patch size, microclimate, degree of isolation (increasing in modern times due to urban expansion), and relative proximity to other patches, so that population stability patch-to-patch and region-to-region is variable. These complex metapopulation dynamics have local function nested within larger patterns of function on ever increasing geographic scales.

In this context, it should be noted that southern California—the range of the QCB—differs substantially from the coastal central Californian domain of Bay Checkerspot in three important ways. First, the southern climate is generally warmer and dryer, with winter precipitation less reliable and highly variable. QCB adults have higher vagility (disperse farther) than Bay Checkerspot. This, along with the QCB hilltopping mating strategy and marked capacity for multiple year larval diapause, enhances QCB fitness in contexts of scarce or poor habitat quality and associated low density adult populations. Secondly, as mentioned above, serpentine geology is absent in the south. Various special soil substrates in southern California support *P. erecta* and present tremendous variation in quality and suitability for the butterfly. Plant size, pubescence, shape, and seasonal longevity, and possibly also nutritional qualities, are partially driven by soil substrate, adding another dimension of uncertainty to quality of butterfly habitat patches. Finally, a portion of the QCB population exists at higher elevations using a suite of other food plants, with these populations having different ecology from the lower elevation populations using *P. erecta* (Pratt et al. 2001). These differences have implications for QCB evolution, biology, ecology, and population dynamics on the whole. QCB adaptations have gone even further in that at higher elevations (to 5,000 ft) in eastern portions of the distribution, populations have been found to use *Plantago patagonica*, *Antirrhinum coulterianum*, and *Collinsia concolor* (all Plantaginaceae); and *Cordylanthus rigidus* (Orobanchaceae) (Pratt et al. 2001). All of the QCB hostplant species are phylogenetically related generally (until recently many of these plants have been classified as Scrophulariaceae), and their familial taxonomy has recently undergone revision resulting from genetic studies (Olmstead et al. 2001). Although *E. editha* are oligophagous (feed upon a limited range of plant species) and feed primarily upon plants contained within the Scrophulariaceae and Plantaginaceae, most local populations tend to be monophagous (feed on only one plant species; White 1974; Scott 1986). Annual precipitation is generally higher at these higher elevations, and these novel food plants are also generally more reliable. The butterfly populations in these areas appear generally at lower densities, but are more stable than those lower elevation populations reliant on *P. erecta*. So at the largest geographic scale across distances spanning 10–50 km, the metapopulation structure incorporates perhaps a relatively steady flux of dispersing females from higher elevations to low, and so occasionally may restart and feed into the boom-bust cycle so prominent with the butterfly-*P. erecta* ecology.

San Francisco Regional Species

The San Francisco Peninsula and the adjacent bay are long, linear features sliced from North America by the actions of the great San Andreas Fault. They tend north-south, parallel to the outer Pacific coast, with peninsular mountains thrust up by the fault with the Bay as an associated area of geological subsidence. Today, the Peninsula (proper) and adjacent bay extend for some 60 km from San Francisco to San Jose. Prehistorically, the southern end of the San Francisco Bay has extended much farther south than it does today and from both climatological and geological perspectives, this peninsula extends for more than a hundred kilometers. The Santa Cruz Mountains form the base of the peninsula and are themselves an ecologically isolated mountain system. Climatological conditions are strongly influenced by oceanic weather generated on the cold northern Pacific Ocean, so that the head of the peninsula, now occupied by the City of San Francisco, bears the brunt of marine winds and fog. Conditions become progressively more temperate to the south and interior (east side) of the peninsula, culminating with hot summer conditions more typical of the inner Coast Ranges. The isolation of this lengthy peninsula together with its gradational insular climates created a hotbed of evolutionary endemism for its butterflies, especially on the most isolated northern tip where special microclimates reach their highest intensity. Although all of its species have wider distributions in western North America, the list of its endemic subspecies include Sthenele Satyr (*Cercyonis sthenele sthenele*, extinct since 1880), Xerces Blue (*Glaucopsyche lygdamus xerces*, extinct since 1943, formerly considered a distinct species), Strobeen's Parnassius (*Parnassius cloudius strohbeeni*, extinct since 1956), Mission Blue (*Icaricia* [*Plebejus*] *icarioides missionensis*, federally listed), San Bruno Elfin (*Incisalia mossii bayensis*, federally listed), Callippe Fritillary (*Speyeria callippe callippe*, federally listed), Unsilvered Fritillary (*Speyeria adiaste adiaste*, proposed for federal listing), and loosely, the Bay Checkerspot (*Euphydryas editha bayensis*, federally listed) (Howe 1975). The geographical, geological, and climatological conditions present in the Bay Area have given rise to many locally adapted subspecies, many of which are now federally listed, primarily as a result of habitat loss of their restricted habitats.

Callippe Fritillary

Adults of the Callippe Fritillary are dark, melanic creatures compared with their conspecifics elsewhere across western North America—a clear adaptation to the relatively cool climate prevailing on the outer end of the San Francisco Peninsula. Their dark coloration allows them to bask and more efficiently elevate body temperatures on summer mornings after the fog has lifted. They inhabit grassland hillsides and swales where their perennial violet host (*Viola pedunculata*) occurs in localized stands. Although the adults are capable of flying long distances of several kilometers, and sometimes do, they tend to show high fidelity to the small habitat patches from which they have developed. Dozens may be found flying about in a small area of a few acres while few if any can be found outside the area. The flight season may extend from late May into July, with the males emerging early, and only gravid females remaining into the latter part of the season. Callippe, like all fritillaries, fly after their food plants have matured, senesced, and dried. Eggs are laid in and around the dry host vegetation. The eggs hatch within several days, and the first instar larvae diapause through the remainder of the year until the following early spring when the food plants renew their annual growth. Larvae then develop through the spring, pupate, and emerge as adults in early summer.

Native grasslands and the *Viola*, along with Callippe Fritillary, were probably once widespread over the San Francisco Peninsula. The ecology was irrevocably changed with the introduction of European grasses (especially *Bromus* and *Avena*) and lands supporting these habitats were all but consumed by the growth and development of San Francisco and other communities along the eastern margin of the peninsula. Today, nominate Callippe is found reliably only on the most prominent of undeveloped remnants, San Bruno Mountain. Other populations of *Speyeria callippe* are found farther south in the Bay Area, perhaps representing clinal intergrades between nominate *S. c. callippe* and more southerly populations of *S. c. comstocki*, although this taxonomy remains to be resolved. Threats to the butterfly did not stop with the protection of San Bruno Mountain as a result of the first Habitat Conservation Plan in 1983, because this protection, along with the lack of disturbance, has allowed both the predominance of dense exotic grasses (which preclude the *Viola*) and longer term successional advancement of shrublands (dominated by native *Baccharis* and exotic gorse, *Ulex europaeus*) to overtake the remaining open grassland habitats that support the butterfly (Longcore et al. 2010). Large-scale reversal of these successional changes will likely require use of fire, selective and limited grazing, or other means of grass and shrub removal.

Mission Blue

Boisduval's Blue (*Icaricia icarioides*) is diverse across its range on western North America, with six subspecies just within the San Francisco Bay area (between Monterey County and Marin County). Larvae feed on various *Lupinus* species. The Mission Blue uses silver lupine (*L. albifrons*), summer lupine (*L. formosus*), and varicolor lupine (*L. variicolor*), which grow in open grassland environments. The bright blue adults fly from late March into June. Larvae, like those of El Segundo Blue, may be tended by ants, although this relationship is not obligatory to the butterfly's development.

The story of the Mission Blue is similar to that of the Callippe Fritillary. Both their habitats, after being decimated by urban growth, continue to be lost to the insidious successional transformation of grasslands to vegetation dominated by *Baccharis* and other woody shrubs. This butterfly persists on a few remaining upland areas such as San Bruno Mountain, Twin Peaks, and the Marin Headlands.

Lange's Metalmark

The San Francisco Bay forms a gap in California's coast range allowing the cool marine air to flow east across the Sacramento-San Joaquin Delta into the Central Valley, and so the Delta is a place of almost incessant winds. Here on the Delta is the final convergence of all the rivers and their tributaries issuing from the western slopes of the Californian Cascade and Sierra Nevada ranges, where their waters mix in the estuaries and make their final run into the Bay and on to the sea. Prehistorically, the rivers ran wide and unchecked in spring flood, carrying with them, and discharging into the valleys and delta, vast loads of sand and silt. In summer the riverbeds dried and the winds blew exposed sands into great dunes that punctuated the arid San Joaquin Valley and its delta.

The Mormon Metalmark Butterfly is another species of broad range and geographic variation through the arid regions of the American West. Numerous subspecies, races, and populations are variously specialized on different species of buckwheat (*Eriogonum*) or are otherwise diversified by geographic isolation. Naked buckwheat (*Eriogonum nudum* var. *auriculatum*) is a small perennial shrub widespread on rocky or sandy substrates through arid portions of the California Coast Ranges, Sierra Nevada, and valley dune environments and in many areas, supports populations of metalmark butterflies. The San Joaquin Delta dunes with their stands of naked buckwheat and resident metalmark populations were especially remote relative to similar populations farther south. Prevailing southeasterly winds from the San Francisco Bay further diminished butterfly movement—and gene flow—in the direction from the more southerly populations to this northernmost outpost on the San Joaquin Delta. In this context, the Lange's Metalmark (*Apodemia mormo langei*) evolved specialized on the naked buckwheat growing in the Antioch Dunes.

The Delta was dredged and drained by canals and levees, the surrounding lands went into agriculture, and the wild rivers feeding the Delta were dammed and their flows moderated. Natural aeolian sands and the processes that generated them no longer augment the remnant habitat at the Antioch Dunes. The decline of Lange's Metalmark is most acutely attributable to extensive mining of dune sands for construction materials used in San Francisco. As the cities of Antioch, Pittsburg, and Oakley grew, their urban developments further subsumed the sand dunes. By the 1970s most of the dunes were gone. The remnant sand dunes became a National Wildlife Refuge in 1980, complete with the protections, fencing, and cessation of any off-road vehicle disturbance, further cementing the process of dune stabilization. Within the several years after sand mining and heavy recreational off-road vehicle uses had stopped, the highly disturbed habitat underwent the early successional stages of vegetation and high populations of both buckwheat and the butterfly resulted for a time (Fig. 2). This boom was fueled by active management and planting of larval food plants at the Refuge. With the absence of active (shifting) dune substrates, however, the soils became increasingly stabilized, especially by exotic grasses (*Bromus*) and forbs. Over time, the exotic vegetation further inhibited growth and recruitment of the naked buckwheat and the successionally prime conditions eventually phased out. Now the butterfly is on the brink of extinction, with its peak numbers in the few dozens since 2006 (Fig. 2).

A remaining few acres of habitat supporting the butterfly is steadily being overrun by evergreen oak, *Baccharis*, and toyon, and yet, because this is the last place of butterfly persistence, Refuge managers are reluctant to cut out the oaks and *Baccharis* for fear that this change may in some unknown way finally push the butterfly over the brink.

Amidst these grim prospects, a captive rearing project has been undertaken and some agency advisers have even contemplated collecting all remaining butterflies into captivity. Wild females have been annually collected for eggs and resulting larvae reared with the hope that a captive population may be established, but captive matings proved preciously difficult to achieve and prospects of long-term success remain unresolved, although captive-reared larvae and adults have been used to augment numbers in the field (Johnson et al. 2011). Community outreach and education programs have generated much interest and volunteer involvement with conservation of the butterfly. In an ambitious effort to restore sand substrates to a large portion of the Refuge now lacking the butterfly, 14,000 m^3 of bay sediments were dredged and pumped onto the Refuge and allowed to dry (Fig. 3). Planned use

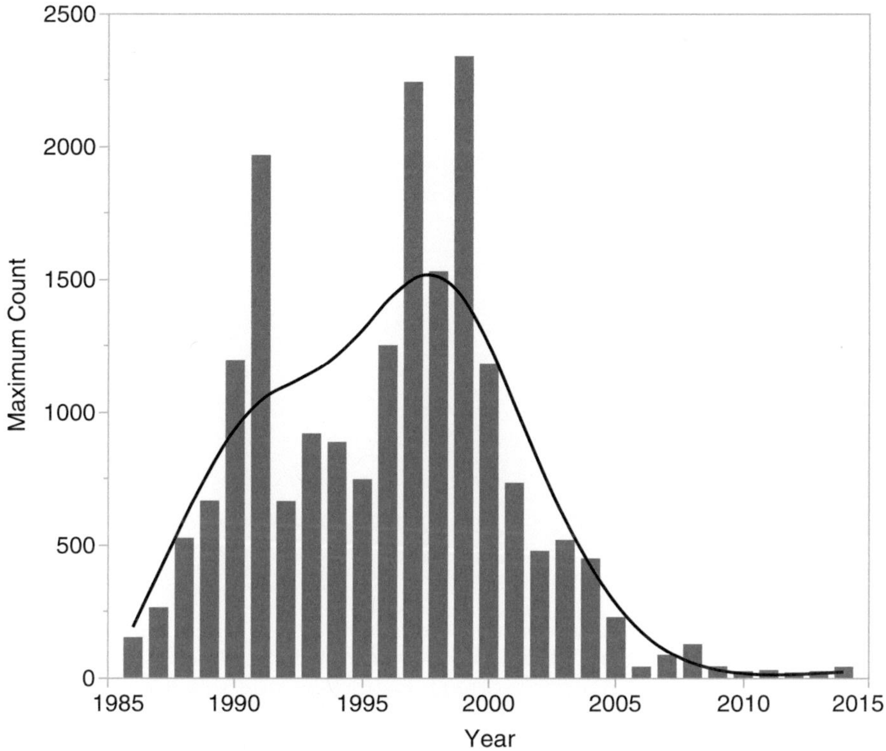

Fig. 2 Maximum daily count of Lange's Metalmark Butterfly at Antioch Dunes National Wildlife Refuge, 1986–2014 (Data from U.S. Fish and Wildlife Service). Smoothed curve fit is from kernel density estimation and shows long-term trend

of a conveyor belt may use winds to winnow the sand out from finer silts and clays. It is hoped that with augmentation of sands vegetational succession may rescue habitat conditions for naked buckwheat, the butterfly, and other federally endangered sand associated plants on the Antioch Dunes National Wildlife Refuge.

Lessons from Conserving California's Endangered Butterflies

The status of the California endangered butterflies varies widely across species. For those with which we have been involved in conservation efforts, a number of themes have emerged over the years. These lessons are not exhaustive or universally applicable, but provide some guidance for those embarking on butterfly conservation efforts in this region and beyond.

Fig. 3 Native early-successional vegetation growing on bay sediments pumped onto the site 1 year earlier at Antioch Dunes National Wildlife Refuge (Photograph: K. H. Osborne)

Cities Matter

Most of traditional conservation biology and indeed of conservation thinking before the advent of conservation biology as a discipline looks at cities as places that are of little interest or consequence. Indeed, major North American conservation initiatives center around connecting large blocks of wilderness with other blocks of wilderness through corridors designed for movement of large mammals (e.g., Noss 1994). Cities received little attention, with the exception of studies that further concentrated on the importance of predators, such as the now-classic study of San Diego canyons (Soulé et al. 1988; Crooks and Soulé 1999). Although butterflies found in remnant habitats in cities were listed under the Endangered Species Act in the 1970s (e.g., El Segundo Blue, Palos Verdes Blue), this did not result in significant acknowledgment of these types of spaces in cities as important reservoirs for biodiversity. The San Diego canyon studies brought the scale of conservation interest down to canyons that might be inhabited by coyotes (and therefore support more native birds), but did not recognize much value for even smaller fragments, or habitats that might be part of the city matrix itself.

Conservation of listed butterfly species cannot succeed without recognizing and embracing cities as places that support significant biodiversity (Eyre et al. 2003; McIntyre 2000; Watts and Larivière 2004). Sometimes noxious commercial and industrial land uses are found adjacent to the habitats of endangered insect species in California and these land uses in fact protect remnant habitats from development (Longcore and Rich 2008). Half of the 14 listed butterfly species in California are found only in habitats in urban areas (Table 1). This does not mean that endangered butterflies are found in the exotic landscaping of the city, but that their habitats, down to quite small fragments (tens of square meters) can in some instances persist and support these species even when surrounded by urbanization.

Natural habitats in cities are important for insect biodiversity for a number of reasons: (1) they contain remnants of habitat types that have always been rare and may have resident butterflies adapted to them (e.g., the El Segundo Dunes in Los Angeles or the coastal bluffs of the San Francisco Bay region), (2) they may by chance contain the remaining fragments of once widespread habitats and the associated resident butterfly species (e.g., Callippe Fritillary as a once widespread coastal subspecies specializing in annual grasslands), or (3) they contain specialized but often small habitats used by widespread species, for which overwintering sites for Monarchs is the prime example (e.g., Ellwood Mesa in Santa Barbara County). California may differ from other areas of North America because of its unique biogeography and high levels of endemism, but as a general rule urban and suburban habitats are more important to invertebrate conservation than is often appreciated.

In recognizing that cities matter to butterfly conservation, explicit and conscious awareness of scale is important. Because the relative meaning of "more isolated" and "smaller" is geared toward vertebrates, e.g., birds (Crooks et al. 2001) or mammals (Bolger et al. 1997), the habitats that may be valuable to insects are often presumed to be ecologically insignificant. Research, however, has illustrated that even small remnants (50–1,000 m^2) isolated from other native vegetation are useful in sustaining populations of invertebrates and can serve as stepping-stone habitats (Abensperg-Traun and Smith 1999). Of course more species are found in large fragments than small, mainly because of presence of food resources more than area, but small fragments of 1–2 ha (and smaller) help support rare butterflies (Rodrigues et al. 1993). For example, a restoration project for El Segundo Blue butterfly in Torrance, California reestablished native coastal dune and bluff plant species, including the food plant *Eriogonum parvifolium*, on a 0.3-ha site isolated from a small remnant population of El Segundo Blue by 360 m (Fig. 4). Without assistance, the butterfly colonized this restored habitat as well as a second 1.5-ha restoration site sandwiched between parking lots and bicycle paths (Schoch 2007) and remains stable there 7 years later (Longcore pers. obs.). Such small habitats should be able to support this endangered species indefinitely with modest (but essential) maintenance, yet they are at a scale that is not usually even recognized as worthwhile by conservation planners. Grizzly bears used to roam the beaches of Los Angeles and humans would never allow them to return. But a dime-sized endangered butterfly is now so common in places that a summer visitor cannot help but to see one flitting about on a walk from the parking lot to the beach.

Table 1 Federally listed Lepidoptera in California, specifying whether its habitat is found in a landscape that is predominantly urban or rural

Species	Habitat context	Larval food plant and habitat	Citations
Hesperiidae			
Pyrgus ruralis lagunae Laguna Mountains Skipper	Rural	*Horkelia clevelandii* in montane meadows	(U.S. Fish and Wildlife Service 1997)
Lycaenidae			
Incisalia mossii bayensis San Bruno Elfin	Urban	*Sedum spathulifolium* on steep north-facing slopes	(Emmel and Ferris 1972)
Lycaeides idas lotis Lotis Blue	Rural, presumed extinct	*Lotus formosissimus* in coastal bogs	(Arnold 1983; U.S. Fish and Wildlife Service 2007)
Euphilotes enoptes smithi Smith's Blue	Rural	Two species of *Eriogonum* on coastal dunes	(Arnold 1983)
Euphilotes battoides allyni El Segundo Blue	Urban; dune habitats at end of airport runway, next to refinery, and long heavily visited urban beach	Seacliff buckwheat (*Eriogonum parvifolium*) on coastal dunes and bluffs	(Mattoni et al. 2001; Schoch 2007)
Glaucopsyche lygdamus palosverdesensis Palos Verdes Blue	Urban/Industrial; on military fuel depot, all sites surrounded by urban development	Locoweed (*Astragalus lonchus trichopodus*) and deerweed (*Acmispson glaber*) in coastal sage scrub; disturbance dependent	(Mattoni 1995)
Icaricia icarioides missionensis Mission Blue	Urban	*Lupinus albifrons*, *L. variicolor*, and *L. formosus* in maritime influenced grassland early-successional coastal scrub	(Arnold 1983; Longcore et al. 2010)
Riodinidae			
Apodemia mormo langei Lange's Metalmark	Urban/Industrial; remnant dunes between power plant and gypsum plant	*Eriogonum nudum* var. *auriculatum* on riverine dunes	(U.S. Fish and Wildlife Service 1984)
Nymphalidae			
Euphydryas editha bayensis Bay Checkerspot	Urban	*Plantago erecta*, *Castilleja densiflorus*, *C. exserta* on serpentine grasslands	(Murphy and Weiss 1988)

(continued)

Table 1 (continued)

Species	Habitat context	Larval foodplant and habitat	Citations
Euphydryas editha quino Quino Checkerspot	Rural; many remaining habitats are in urbanizing areas	Several, predominantly *Plantago erecta*	(Mattoni et al. 1997)
Speyeria callippe callippe Callippe Fritillary	Urban	*Viola pedunculata* in grasslands	(Longcore et al. 2010)
Speyeria zerene behrensii Behren's Fritillary	Rural	*Viola adunca*	(U.S. Fish and Wildlife Service 2003a)
Speyeria zerene hippolyta Oregon Fritillary	Rural	*Viola adunca*	(McCorkle and Hammond 1988)
Speyeria zerene myrtleae Myrtle's Fritillary	Rural	*Viola adunca*	(Launer et al. 1992)

Fig. 4 Example of habitat restoration for El Segundo Blue butterfly in Torrance and Redondo Beach, California. Native shrubs were established with a cover crop of fescue (*top*) and matured to a diverse community of strand and bluff species that was colonized by the butterfly (*bottom*) (Photographs: Travis Longcore)

Connectivity Does Not Always Mean Corridors

Providing connectivity is perhaps the most conceptually simple conservation principle and the public and conservation advocates easily grasp it. As a result, we have found that connectivity as defined by a terrestrial corridor between habitats becomes a criterion for judging the value of places without consideration of what species might use such a corridor. Often, we have encountered planners and funders who conclude that a place that is not connected to other habitats by a land-based contiguous corridor has no conservation value. Not all species, however, require a contiguous land-based corridor to move between potential habitats (Longcore 2006). To the contrary, how "connected" a particular natural area is depends entirely on what species is considered.

To be sure, for butterflies as a whole, inter-patch distance affects colonization and population persistence (Hanski et al. 1996; Hanski 1999; Bergman and Landin 2002) and corridors can enhance movement between habitat patches (Haddad 1999; Sutcliffe and Thomas 1996). Dispersal ability varies within groups, however, and it is not uncommon to have butterfly species that regularly cross two-lane highways or traverse the entire urban landscape, while for others movement across such barriers is limited (Munguira and Thomas 1992). Some investigators have concluded that stepping stones of suitable habitat may be more effective than corridors in promoting inter-patch dispersal (Schultz 1998). Dispersal over inhospitable habitat for habitat specialist species may be restricted to 1 km or less, while others can locate specialized habitats nearly anywhere within an urban matrix. Hence, conservation strategies can focus on protection of suitable habitat areas, even at a very small scale. This may include habitat elements located outside of recognized natural areas that consist only of appropriate foodplants, such as the Green Hairstreak Corridor project in San Francisco (Fig. 5) (see Nature in the City; http://www.natureinthecity.org/).

Next, because of the nature of butterflies and our ability to relocate them relatively easily, humans can provide connectivity for rare butterflies in ways that would be difficult to accomplish with large predators. Once any regulatory considerations are resolved, it is not expensive to move butterflies between small habitats in cities and this may indeed be the most effective long-term conservation strategy. It could be coupled with a captive breeding program, such as our work with Palos Verdes Blue where a captive breeding program provides stock to recolonize isolated protected areas on the Palos Verdes Peninsula (Finnegan 2010). Of course, some species have extraordinarily specialized habitat requirements that might include mutualistic interactions with ants and which may or may not be easy to replicate even if human-assisted colonization is provided.

Fig. 5 Project description from Nature in the City for its Green Hairstreak Corridor project (Reprinted from http://www.natureinthecity.org/index.php/programs/the-green-hairstreak-project). Remnant natural areas are stitched together with plantings of larval foodplants

Disturbance Can Be Needed—Why "Fence It and Forget It" Does Not Work

The overarching theme of wilderness conservation is, rightly, that large blocks of habitats be protected from human disturbance. In the past 30 years large-scale natural land management has shifted to recognize and incorporate the role of natural disturbance in protected lands (Sousa 1984; Pickett and White 1985). But even as the role of disturbance is recognized as essential to healthy functioning of large protected lands, a fear of all human disturbances permeates management of natural habitats in urban contexts. This is problematic because many rare and endangered butterfly species depend on larval foodplants that in turn depend on various natural disturbances for their long-term persistence and health (Table 1). Urban fragments, for various reasons, may no longer experience natural disturbances and so the standard conservation approach of minimizing human disturbances ("fence it and forget it") has the result of degrading habitat for those butterflies that depend on early-successional foodplants. Several examples illustrate this important point.

Palos Verdes Blue butterfly has two known larval foodplants, locoweed and deerweed. Both are in the Fabaceae and grow best following land disturbances such as fire, landslides, or at the edge of wildlife trails. Both are perennials, but are short-lived and prone to senescence after several years. The Defense Fuel Support Point, San Pedro was closed to public access from its construction during World War II until the mid-1990s. When it was opened to scientists to survey for insects, an unknown population of Palos Verdes blue butterfly was found there (Mattoni 1995), at which point regular surveys were started first of the adult butterflies and later of the habitat elements (Mattoni and Longcore 2002; Longcore and Osborne 2013). Substantial restoration efforts to establish coastal sage scrub (including the larval foodplants) have been undertaken since 1994. By the mid-2000s it became clear that the restoration for coastal sage scrub was becoming "too successful" in that early-successional habitats with foodplants for Palos Verdes blue were being replaced by more mature scrub without the larval foodplants (Osborne 2002a; Longcore 2007a, b). Review of historical aerial photographs provided evidence that disturbance from operations at DFSP had been much greater before the discovery of the butterfly in 1994 than afterwards. A comparison of photographs from 1994 to 2009 shows succession and closure of canopy in a large area in the central valley of the installation during this period, as well as succession in other locations that had supported the butterfly during the 1990s (Longcore and Osborne 2010).

The importance of artificial disturbance to the ongoing presence of larval foodplant at DFSP was confirmed when a large area of sandy soil was graded for purposes other than habitat management. Both the flat pad and the raw slope adjacent to it were quickly colonized by deerweed. Captive-bred PVB were then released at the site and their offspring were observed for several seasons thereafter (Longcore and Osborne 2012). With this information and at our urging, a pilot project was initiated to mechanically disturb an acre of exotic grassland next to a stand of deerweed. The top layer of soil, along with the invasive European grasses in it, was removed, leaving bare soil. After a year, it was evident that deerweed was establishing itself (Fig. 6) and in subsequent years will become a location for butterfly recolonization efforts.

Mission Blue butterfly (*Icaricia icarioides missionensis*) is a small blue butterfly in the Lycaenidae that has a single flight period in the spring and is found in the hills stretching north and south of San Francisco. Like all of the Boisduval's Blue subspecies, Mission Blue specializes on lupine plants as a larval foodplant (Downey 1962), with three perennial lupines (*Lupinus albifrons* var. *collinus*, *L. formosus* var. *formosus*, and *L. variicolor*) used by this subspecies. These foodplants are characteristic of early-successional habitats, as are most lupine species. The dynamics of mission blue at the Golden Gate National Recreation Area (GGNRA; MacDonald et al. 2012) and the San Bruno Mountain Habitat Conservation Plan (Longcore et al. 2010) reserves probably depend at least in part disturbance. In

Fig. 6 *Top*: Experimental disturbance site at Defense Fuel Support Point, San Pedro. The top layer of soil and associated cover of nonnative grasses was removed. *Bottom Left*: Disturbance experiment site a year later. The scraped area has many annual weeds, but also seedlings of the perennial deerweed (*Acmispon glaber*). *Bottom Right*: Deerweed seedlings in disturbance experiment area. Seedlings are to the *right* and *left* of the pen, surrounded by nonnative *Erodium* plants, but no invasive grasses (*Top photograph*: Palos Verdes Peninsula Land Conservancy. *Bottom photographs*: K. H. Osborne)

a review of population monitoring, MacDonald et al. (2012) strongly recommend the use of intentional disturbance as a management tool for habitat enhancement and restoration at Milagra Ridge and Marin Headlands in GGNRA. Such methods presumably could include controlled burning, which is not always an alternative because of air quality regulations, or other approaches. Disturbance effectively ensures regeneration of suitable foodplant patches (Wayne et al. 2009). At San Bruno Mountain, higher densities of lupines and butterflies have been reported on cut slopes and roadcuts; while these provide early-successional habitat, they frequently are in more sheltered and wind-protected locations also (Thomas Reid Associates 1999). Failure to provide disturbance/vegetation management to keep these early-successional habitats is likely to threaten the viability of the listed species (Longcore et al. 2010). This presents a challenge that is often faced in California butterfly conservation—resistance to removing native vegetation (e.g., scrub encroaching on grasslands or late succession scrub species replacing early-succession scrub species) by both agencies and advocates hesitant to remove any native plant.

Such California examples are consistent with other butterflies with early-succession larval foodplants in other regions. Availability of recently burned high-quality early-successional habitat was shown to be very important for Karner Blue butterfly in Ohio; number of oviposition events within managed (burned or mowed) grassland was significantly larger (122 of 127 events) than in habitats not managed in this manner (Pickens and Root 2009). For Miami Blue butterfly in Florida, hurricanes produce the disturbance conditions that promote its foodplants, including nickerbean (Daniels et al. 2002) and blackbead (Cannon 2007). A large stand of blackbead at Bahia Honda State Park in the 2000s was the result of disturbance from Hurricane Georges in 1998 (Daniels et al. 2002) and sites discovered to be occupied by Miami Blue at Boca Grande and the Marquesas Islands had been "transformed" by hurricane Wilma 1 year prior (Cannon 2007).

Disturbance can also be important where foodplants are not necessarily early-succession specialists, but where they are being excluded by exotic species through direct competition. For example, the *Plantago* foodplants used by Bay Checkerspots grow better and butterflies thrive when cows graze annual exotic grasses (Weiss 1999). Lack of disturbance from shifting sands is one mechanism that appears to be tipping the scales in favor of exotic plant species at the Antioch Dunes National Wildlife Refuge, with resulting adverse consequences for the endangered Lange's Metalmark. The metalmark's obligate foodplant, naked buckwheat (*Eriogonum nudum* var. *auriculatum*) (U.S. Fish and Wildlife Service 1984), does not recruit in the dense thatch of invasive annual grasses. The dunes are greatly depleted as a result of sand mining so refuge managers are building a new dune from dredged sand on the premise that the Aeolian disturbance of the new dune will favor the buckwheat and other endemic dune plants protected at the refuge (Fig. 3).

In short, human-created disturbances or intentional and directed use of natural disturbances will be necessary to maintain habitat for many endangered butterflies (Morris et al. 1994). These efforts may come into conflict with management of habitat for later succession species and resistance from land managers or a public accustomed to a hands-off approach to conservation management.

Numbers Matter, But Individuals Are Not Sacred

Conservation of small-bodied, short-lived species with high reproductive capacity is fundamentally different from the typical flagstone conservation example like grizzly bears or California Condors. Unlike these species, with their small number of offspring, butterflies have the potential for hundreds and thousands of progeny, only a tiny fraction of which reach reproductive age. This difference has a number of consequences for thinking about and assessing population numbers and the conservation value of individuals of the species.

Although it is established in the literature, it bears emphasizing that the approach to a population viability analysis (PVA) for butterflies is quite different than for large, long-lived species. Large-species population assessments concentrate on enumeration of each individual and discussion focuses around whether the population size is sufficiently large to avoid demographic or genetic disasters (Lande 1995; Interagency Grizzly Bear Study Team 2005; Snyder and Snyder 2000; Snyder and Schmitt 2002). In contrast, PVAs for butterflies (should) draw on metapopulation theory to incorporate interacting populations within a spatially explicit context (Schtickzelle et al. 2005; Radchuk et al. 2012). Non-spatial PVAs can be and are used for individual populations (Schultz and Hammond 2003), but they face certain difficulties, such as how to address multi-year diapause when addressing extinction thresholds (Longcore and Osborne 2013).

The implication of using PVA analyses is that population sizes can be measured with some degree of accuracy and that the number of individuals is an important predictor of survival. In many ways, however, this does not hold true for butterflies. It is not simple to estimate total brood size for butterflies (Zonneveld 1991; Mattoni et al. 2001; Zonneveld et al. 2003). Butterflies have the potential for explosive population growth, and it is difficult to have the historical data available to calculate the growth rate (lambda) of the species accurately and to incorporate this into a PVA. We have been calculating the extinction risk for the Palos Verdes blue for a decade with a diffusion/approximation model (Morris et al. 1999) and each year have to update lambda as a result of the most recent year of surveys (Longcore and Osborne 2013).

In many respects, population numbers are the wrong things to focus on when assessing the status of a rare butterfly. Rather, the key question is how much habitat exists and is occupied. For Palos Verdes Blue, we calculated the percent of transect segments occupied by adult butterflies each year, and this metric of geographic spread gave a signal of decline before it was apparent in the population estimates (Fig. 1). For Mission blue, we used a grid of 250-m cells across San Bruno Mountain (Longcore et al. 2010) and Golden Gate National Recreation Area

(MacDonald et al. 2012) to track shifts in the occupancy pattern. This approach was developed in part because of the limitations of the previous monitoring efforts to develop population estimates, but in many respects occupancy is the more important question. Butterfly numbers can and do fluctuate wildly with weather conditions (Mattoni et al. 1997) and from a conservation perspective, having a widely distributed population of a rare species is better then having a large number concentrated in a single location. Indeed, Schultz and Hammond (2003) demonstrated that extinction risk for Fender's Blue decreased more with additional populations than with increasing population sizes at existing sites.

Small population sizes still matter for butterflies in terms of demographic (Calabrese and Fagan 2004) and genetic (Saccheri et al. 1998) impacts, but conservationists would be well advised to pay attention to how much area is occupied before concentrating on those numbers.

A corollary of understanding the relative importance of occupancy over population size is that individual butterflies, and especially individual eggs and larvae, are not sacred. That is, within the context of the high mortality experienced through the developmental process and the large reproductive capability of females to lay hundreds of eggs, an individual butterfly is far less important to the survival of the species than is the existence of the appropriate habitat (foodplant, nectar resources, etc.), so long as some butterflies are present to use that habitat. It should be considered an acceptable and routine risk to allow some incidental take of threatened and endangered butterflies in the process of habitat restoration and management. It is simply not equivalent to killing a grizzly bear or a California Condor, even if regulatory agency staff seem capable of seeing it that way. To the contrary, efforts to avoid incidental take by forcing avoidance of vegetation management where a single larva or pupa might be disturbed has probably done more harm than good for recovery of listed butterflies. It is far more important that appropriate disturbance be allowed to promote habitat for those species with transient post-succession foodplants than it is to avoid all mortality of every pupa or larva. It is furthermore ill advised in habitat restoration efforts to attempt strict avoidance of incidental take, such as salvage of pupae, unless those individuals are easy to find. For Palos Verdes Blue, a pre-development pupal salvage effort turned out to cost tens of thousands of dollars and yielded two pupae (Longcore et al. 2005).

Our admonition to accurately assess the relative benefits of habitat management against the possible incidental take of a few individuals of a species should not be interpreted as an argument that "take" is not important. Collectors can and do take large numbers of specimens, either legally or illegally, and even a small number of specimens of an extremely rare species could be consequential. But when it comes to the possible incidental take of eggs or larvae in the course of necessary habitat management, the risks of not managing may, in many instances, outweigh those of taking the action.

Ranges Do Not Collapse to Highest Quality Habitat

One of the challenges of working with the many butterfly species that persist only on remnant habitats in now-urban areas is that it is easy to assume that those sites are examples of high-quality habitat for the species. The rare butterflies of coastal (and largely urban) California, however, may be limited in range to locations that may not represent the best conditions for the species (or subspecies) historically. As a consequence, both the habitat characteristics and the processes that maintained them may not be immediately obvious without considering a history of the species that might extend well beyond its remnant distribution.

As discussed above, Quino Checkerspot Butterfly was once distributed over most of coastal southern California and southward into Baja California (Mattoni et al. 1997). Its most common larval foodplant across the forb and grass dominated coastal plains and hills, *Plantago erecta*, would have decreased in distribution significantly following the introduction of Mediterranean grasses and widespread grazing by cattle and sheep during the Mission Era. The abundance of Quino Checkerspots would have been breathtaking during years with sufficient rainfall, making its decline and range collapse akin to the loss of the Passenger Pigeon (Mattoni et al. 1997)—an immeasurably numerous species declining to extreme rarity. The more inland populations known and discovered around the time of listing in 1997 and more recently discovered higher elevations of Quino Checkerspot (Pratt et al. 2001) probably do not represent the typical habitat of the heart of the species' historical distribution along the coast. Some of these habitats have vanished entirely, but leave evidence that they had appropriate foodplants for the butterfly (Mattoni and Longcore 1997). Recognizing these historical conditions is important to define restoration objectives and to ensure that recovery is contemplated for places where even the memory of the butterfly's occurrence has faded through generations of shifting baselines (Papworth et al. 2009).

That historical habitat may have been quite different from current habitat is also important when developing species distribution models, especially those from presence data (Elith et al. 2011; Philips et al. 2006). Presence-only distribution models for rare species have the potential to emphasize conditions at those remnant sites and miss the possibilities for restoration and recovery within the former range (MacDonald et al. 2010). Because butterflies can be reintroduced to fragments of habitat within a developed landscape (e.g., Mission Blue in San Francisco; Weiss et al. 2012), historical distribution of species should always be considered when defining the landscape position and characteristics of suitable habitat.

Scale Is Different—A Molehill Can Be a Mountain

Humans perceive the world and operate on the scale of our bodies. Even though we intellectually may understand the vast difference in scale between the size and weight of a butterfly and our frame of reference, it can be difficult to incorporate into thinking about conservation. A final lesson that emerges from efforts to understand and conserve California butterflies is how important topographic differences can be, and especially at the scale of butterflies and their larvae.

The considerable and extensive research on Bay Checkerspot around the San Francisco region illustrates how topography affects larvae (Murphy and Weiss 1988; Weiss et al. 1988). Researchers documented that larval distribution was highly clumped and moved according to weather patterns and microclimatic factors. The developmental biology of Bay Checkerspot is a race against the maturation and death of their food plants (White 1974; Weiss et al. 1988). Larvae emerging from diapause in the spring must mature, find mates, and lay eggs, and those eggs must mature and hatch, and those new larvae must then feed at least through two molts before food plant senescence in April or early May. The post-diapause larvae crawl around on the ground consuming hundreds of the tiny food plants. Black coloration of the larvae allows them to bask and warm in the sun, accelerating their food digestion and metabolism (Weiss et al. 1988). Larvae finding themselves on northerly slopes tend to migrate to southerly exposures with more sunlight (Weiss et al. 1987), better basking, and faster growing food plants. Larvae quickly finish their development, pupate, and several days later emerge as adults, flying mostly in March and April.

All of this holds true for Quino Checkerspot as well. These complex patterns of adult emergence, oviposition site selection, and foodplant senescence with the summer drought illustrate the importance of maintaining habitats with many microclimates to persistence of this species. Solar insolation on hillsides (determined in part by topography) where the larvae live affects both the rate of host development and that of the larvae (White 1974; Weiss et al. 1988). In the race against host senescence, post-diapause larvae seek microclimates with high solar insolation to bask (Weiss et al. 1988; Osborne and Redak 2000) and particularly with Quino Checkerspot, the presence of woodland and scrub vegetation in its habitats increases the environmental heterogeneity on a small scale of meters (Osborne and Redak 2000). Microtopographic and microclimatic elements such as these are found at a scale that is easy to ignore and also reinforce the role of disturbance in creating fine scale habitat variability.

El Segundo Blue adult butterflies appear to prefer and exploit microclimates created by dune scrub. Their coastal habitats face persistent onshore winds and butterflies are often observed sheltering in the lee of the dune (i.e., the backdune) (Mattoni 1992; Mattoni et al. 2001) and in the lee of foodplants when present on bluffs or foredunes (Osborne 2001).

Finally, many butterfly species use topographic relief as a means to locate mates. Known as hilltopping (Shields 1967), this phenomenon has been debated in terms of its importance (Ehrlich and Wheye 1988; Baughman et al. 1990; Singer and Thomas 1992), but the conclusion remains that hilltops appear to contribute to population persistence for some species (e.g., Quino Checkerspot; Osborne 2002b; U.S. Fish and Wildlife Service 2003b). Quino Checkerspot mating activity occurs in or near meadows, clearings, or open areas on slopes and ridgelines inhabited by the food plants where the larvae previously developed. Apparently unlike the

Bay Checkerspot, butterflies mate on open or sparsely vegetated hilltops in absence of hostplants. Adult male Quino are found on hilltops (usually only one or two per hilltop) where they exhibit "territorial behavior" consisting of flying sorties from various perches to chase other butterflies, including conspecifics (Sheilds 1967). These males often chase each other high into the air, only to return to different parts of the hilltop. Hilltopping, where male butterflies await the arrival of unmated females on open hilltops, is common in many species of butterflies and the behavior in Quino Checkerspot is well known among experienced southern California lepidopterists (Osborne 2002b). When Quino Checkerspot adult densities are relatively low, mating success derived from facultative hilltoping behavior may be critical to long-term population viability (Osborne 2002b). In some instances these "hilltops" can simply be the highest point in an area and not immediately recognized as a resource, but which become evident if one consciously scales perception of the landscape to that of a butterfly.

Summary and Key Points

Conservation of butterflies, especially in the geographically and biologically diverse context of California, presents a set of challenges that differs considerably from those facing large-bodied, long-lived mammals. The landscape has many locally unique populations of widespread butterflies, often with early-successional habitat affinities, some of which have been recognized and afforded legal protection. As a result, the rules of thumb and first-choice conservation actions are dramatically different:

- Cities matter, because remnant habitats of geographically constrained species can persist in them if appropriate habitat is present, sometimes on quite small plots of land;
- Connectivity between populations may be important but can be provided by stepping stones or human-assisted translocation when necessary, and the lack of a contiguous land-based corridor between habitats does not preclude conservation;
- Disturbance is often essential to maintain butterfly habitat because the dependence of many species on early-successional foodplants means that habitat cannot simply be restored but must be created over and over again;
- Numbers of adult butterflies can fluctuate considerably, so tracking occupancy may be more relevant than population size and loss of some early stage individuals (eggs and larvae) may often be justified in the interest of maintaining appropriate habitat;
- Remnant habitats are not necessarily those that historically were the best for a taxon, so habitat modeling, restoration, and reintroduction efforts should consider past range extents; and
- The whole scale at which butterflies perceive and respond to the physical world is smaller than that typically perceived by humans, so attention to microclimatic and topographic diversity is centrally important to conserving butterfly habitat.

Acknowledgements Our research on Palos Verdes Blue at Defense Fuel Support Point, San Pedro was funded by the Defense Logistics Agency on behalf of the Defense Fuels Region—West. The Central Valley Project Improvement Act (CVPIA) Habitat Restoration Program funded our research on Lange's Metalmark. We thank C. Nagano, A. Owen, and C. Rich for constructive comments on drafts of this chapter.

References

Abensperg-Traun M, Smith GT (1999) How small is too small for small animals? Four terrestrial arthropod species in different-sized remnant woodlands in agricultural Western Australia. Biodivers Conserv 8(5):709–726

Arnold RA (1983) Ecological studies of six endangered butterflies (Lepidoptera, Lycaenidae): island biogeography, patch dynamics, and the design of habitat preserves. Univ Calif Publ Entomol 99:1–161

Axelrod DI (1978) The origin of coastal sage vegetation, Alta and Baja California. Am J Bot 65(10):1117–1131

Baughman JF, Murphy DD, Ehrlich PR (1990) A reexamination of hilltopping in *Euphydryas editha*. Oecologia 83(2):259–260

Bergman K-O, Landin J (2002) Population structure and movements of a threatened butterfly (*Lopinga achine*) in a fragmented landscape in Sweden. Biol Conserv 108(3):361–369

Bolger DT, Alberts AC, Sauvajot RM, Potenza P, McCalvin C, Tran D, Mazzoni S, Soulé ME (1997) Response of rodents to habitat fragmentation in coastal southern California. Ecol Appl 7:552–563

Calabrese JM, Fagan WF (2004) Lost in time, lonely, and single: reproductive asynchrony and the Allee effect. Am Nat 164(1):25–37

Cannon P (2007) Miami blue: rarest of the rare. Keys Sunday, pp 16–18

Crooks KR, Soulé ME (1999) Mesopredator release and avifaunal extinctions in a fragmented system. Nature 400(6744):563–566

Crooks KR, Suarez AV, Bolger DT, Soule ME (2001) Extinction and colonization of birds on habitat islands. Conserv Biol 15(1):159–172

Daniels JC, Pence JA, Emmel TC (2002) Additional records of *Cyclargus ammon* (Lycaenidae: Lycaeninae) in the Florida Keys. J Lepid Soc 60:97–99

Dobkin DS, Olivieri S, Ehrlich PR (1987) Rainfall and the interaction of microclimate with larval resources in the population dynamics of checkerspot butterflies (*Euphydryas editha*) inhabiting serpentine grassland. Oecologia 71:161–176

Downey JC (1962) Host–plant relations as data for butterfly classification. Syst Zool 11(4):150–159

Ehrlich PR, Hanski I (eds) (2004) On the wings of checkerspots: model system for population biology. Oxford University Press, Oxford

Ehrlich PR, Wheye D (1988) Hilltopping checkerspot butterflies revisited. Am Nat 132(3):460–461

Ehrlich PR, Murphy DD, Singer MC, Sherwood CB, White RR, Brown IL (1980) Extinction, reduction, stability and increase: the responses of checkerspot butterfly (*Euphydryas*) populations to the California drought. Oecologia 46(1):101–105

Elith J, Phillips SJ, Hastie T, Dudík M, Chee YE, Yates CJ (2011) A statistical explanation of MaxEnt for ecologists. Divers Distrib 17:43–57

Emmel TC (ed) (1998) Systematics of western North American butterflies. Mariposa Press, Gainesville

Emmel TC, Emmel JF (1973) The butterflies of southern California. Nat Hist Mus Los Angel County Sci Ser 26:1–148

Emmel JF, Ferris CD (1972) The biology of *Callophrys* (*Incisalia*) *fotis bayensis* (Lycaenidae). J Res Lepid 26(4):237–244

Eyre MD, Luff ML, Woodward JC (2003) Beetles (Coleoptera) on brownfield sites in England: an important conservation resource? J Insect Conserv 7:223–231

Finnegan M (2010) A rare butterfly takes flight on the Palos Verdes Peninsula. Los Angeles Times, March 7, 2010

Garth JS, Tilden JW (1986) California butterflies. University of California Press, Berkeley

Haddad N (1999) Corridor length and patch colonization by a butterfly, *Junonia coenia*. Conserv Biol 14:738–745

Hanski I (1999) Metapopulation dynamics. Oxford University Press, Oxford

Hanski I, Moilanen A, Pakkala T, Kuussaari M (1996) The quantitative incidence function model and persistence of an endangered butterfly metapopulation. Conserv Biol 10(2):578–590

Herrmann C (1999) El Segundo blue butterfly (*Euphilotes bernardino allyni*) Malaga Cove food plant population assessment and pilot recovery plan. UCLA Department of Geography, Los Angeles

Howe WH (1975) The butterflies of North America. Doubleday and Company, Garden City

Interagency Grizzly Bear Study Team (2005) Reassessing methods to estimate population size and sustainable mortality limits for the Yellowstone grizzly bear. Interagency Grizzly Bear Study Team, U.S. Geological Survey, Northern Rocky Mountain Science Center, Montana State University, Bozeman

Johnson JJ, Longcore T, Clause A, Lansing A, Renner D, Romero A, Messerschmidt B, Pickett B, Kores P, Courtney K (2010) Captive rearing of Laguna Mountains skipper (*Pyrgus ruralis lagunae*): final report. The Urban Wildlands Group (U.S. Fish and Wildlife Service Cooperative Agreement No. 814307J024), Los Angeles

Johnson JJ, Jones J, Baudour M, Wagner M, Flannery D, Werner D, Holden C, Virun K, Wilson T, Delijani J, Delijani J, Newton B, Gundell D, Thummar B, Blakey C, Walsh K, Sweeny Q, Ware T, Adams A, Taylor C, Longcore T (2011) Captive rearing of Lange's metalmark butterfly, 2010–2011. The Urban Wildlands Group, Los Angeles

Lande R (1995) Mutation and conservation. Conserv Biol 9(4):782–791

Launer AE, Murphy DD, Hoekstra JM, Sparrow HR (1992) The endangered Myrtle's silverspot butterfly: present status and initial conservation planning. J Res Lepid 31(1–2):132–146

Longcore T (2006) The Green Visions Plan for 21st Century Southern California: a guide for habitat conservation, watershed health, and recreational open space. 8. Conservation of biodiversity in the city: an assessment of MRCA projects in the upper Los Angeles River watershed. University of Southern California Center for Sustainable Cities, Los Angeles

Longcore T (2007a) Final report for 2006 Palos Verdes blue butterfly adult surveys on Defense Fuel Support Point, San Pedro, California. The Urban Wildlands Group (Defense Logistics Agency Agreement # N68711-06-LT-R0011), Los Angeles

Longcore T (2007b) Final report for 2007 Palos Verdes blue butterfly adult surveys on Defense Fuel Support Point, San Pedro, California. The Urban Wildlands Group (Defense Logistics Agency Agreement # N68711-07-LT-R0011), Los Angeles

Longcore T, Osborne KH (2010) Final report for 2010 Palos Verdes Blue butterfly adult surveys on Defense Fuel Support Point, San Pedro, California. The Urban Wildlands Group (NAVFAC SW Agreement # N62473-08-LT-R0011), Los Angeles

Longcore T, Osborne KH (2012) Final report for 2012 Palos Verdes Blue butterfly adult surveys on Defense Fuel Support Point, San Pedro, California. The Urban Wildlands Group (Defense Logistics Agency Agreement # N62473-12-2-2101), Los Angeles

Longcore T, Osborne KH (2013) Final report for 2013 Palos Verdes Blue butterfly adult surveys on Defense Fuel Support Point, San Pedro, California. The Urban Wildlands Group (Defense Logistics Agency Agreement # N62473-12-2-2101), Los Angeles

Longcore T, Rich C (2008) Invertebrate conservation at the gates of hell. Wings 31(1):9–13

Longcore T, Mattoni R, Mattoni A (2005) Salvage of individual pupae as a mitigation measure for loss of Palos Verdes blue butterfly habitat. J Lepid Soc 59(2):116–117

Longcore T, Lam CS, Kobernus P, Polk E, Wilson JP (2010) Extracting useful data from imperfect monitoring schemes: endangered butterflies at San Bruno Mountain, San Mateo County, California (1982–2000) and implications for habitat management. J Insect Conserv 14:335–346

Longcore T, Johnson J, Osborne K (2014) Laguna Mountains Skipper captive rearing interim report. The Urban Wildlands Group (National Fish and Wildlife Foundation Agreement No.: #8006.10.033384), Los Angeles

MacDonald B, Longcore T, Dark S (2010) Habitat suitability modeling for Western Snowy Plover in Central California. The Urban Wildlands Group, Los Angeles

MacDonald B, Longcore T, Weiss SB (2012) Status and variability of Mission Blue butterfly populations at Milagra Ridge, Marin Headlands, and Oakwood Valley. The Urban Wildlands Group, Los Angeles

Mattoni RHT (1992) The endangered El Segundo Blue butterfly. J Res Lepid 29(4):277–304

Mattoni RHT (1993) The El Segundo Blue, *Euphilotes bernardino allyni* (Shields). In: New TR (ed) Conservation biology of Lycaenidae (Butterflies). Occasional Paper of the IUCN Species Survival Commission, No. 8, pp 133–134

Mattoni RHT (1995) Rediscovery of the endangered Palos Verdes blue butterfly, *Glaucopsyche lygdamus palosverdesensis* Perkins and Emmel (Lycaenidae). J Res Lepid 31(3–4):180–194

Mattoni R, Longcore TR (1997) The Los Angeles coastal prairie, a vanished community. Crossosoma 23(2):71–102

Mattoni R, Longcore T (2002) Census results for Palos Verdes Blue butterfly and associated species, 1994–2001. In: Mattoni R (ed) Status and trends: habitat restoration and the endangered Palos Verdes Blue butterfly at the Defense Fuel Support Point, San Pedro, California, 1994–2001. The Urban Wildlands Group, Los Angeles, pp 2–10

Mattoni RHT, Pratt GF, Longcore TR, Emmel JF, George JN (1997) The endangered Quino Checkerspot butterfly, *Euphydryas editha quino* (Lepidoptera: Nymphalidae). J Res Lepid 34:99–118

Mattoni R, Longcore T, Krenova Z, Lipman A (1998) Mass rearing the endangered Palos Verdes Blue butterfly (*Glaucopsyche lygdamus palosverdesensis*: Lycaenidae). J Res Lepid 37(1):55–67

Mattoni R, Longcore T, Zonneveld C, Novotny V (2001) Analysis of transect counts to monitor population size in endangered insects: the case of the El Segundo Blue butterfly, *Euphilotes bernardino allyni*. J Insect Conserv 5(3):197–206

McCorkle DV, Hammond PC (1988) Biology of *Speyeria zerene hippolyta* (Nymphalidae) in a marine-modified environment. J Lepid Soc 42(3):184–195

McIntyre NE (2000) The ecology of urban arthropods: a review and a call to action. Ann Entomol Soc Am 93:825–835

Morris MG, Thomas JA, Ward LK, Snazell RG, Pywell RF, Stevenson MJ, Webb NR (1994) Re-creation of early-successional stages for threatened butterflies: an ecological engineering approach. J Environ Manag 42(2):119–135

Morris W, Doak D, Groom M, Kareiva P, Fierberg J, Gerber L, Murphy P, Thomson D (1999) A practical handbook for population viability analysis. The Nature Conservancy, Arlington

Munguira ML, Thomas JA (1992) Use of road verges by butterfly and burnet populations, and the effect of roads on adult dispersal and mortality. J Appl Ecol 29(2):316–329

Murphy DD, Weiss SB (1988) Ecological studies and the conservation of the Bay Checkerspot butterfly, *Euphydryas editha bayensis*. Biol Conserv 46:183–200

Noss R (1994) The Wildlands Project: land conservation strategy. In: Grumbine RE (ed) Environmental policy and biodiversity. Island Press, Washington, DC, pp 233–266

Olmstead RG, dePamphilis CW, Wolfe AD, Young ND, Elisons WJ, Reeves PA (2001) Disintegration of the Scrophulariaceae. Am J Bot 88(2):348–361

Orsak LJ (1988) Synopsis and status of rare, threatened/endangered and lost Orange County butterflies. In: Koerper HC (ed) The natural and social sciences of Orange County. Natural History Foundation of Orange County, Newport Beach, pp 47–61

Osborne KH (2001) Habitat assessment and focused surveys for two endangered butterflies, Palos Verdes Blue butterfly (*Glaucopsyche lygdamus palosverdesensis*) and El Segundo Blue butterfly (*Euphilotes battoides allyni*) on sites at Long Point and Upper Point Vincente, Rancho Palos Verdes, Los Angeles. County Osborne Biological Consulting, Riverside

Osborne KH (2002a) Population monitoring for the Palos Verdes Blue on the Defense Fuel Support Point, San Pedro, California: year 2002 survey results. Osborne Biological Consulting, Riverside

Osborne KH (2002b) Habitat assessment and adult survey for Quino checkerspot butterfly (*Euphydryas editha quino*) on the Metropolitan Water District of Southern California – Lake Skinner Operations Area, Riverside County, California. Osborne Biological Consulting, Riverside

Osborne KH (2003) Adult survey for Laguna Mountain Skipper (*Pyrgus ruralis lagunae*) on the Laguna Mountains, Palomar Mountains, and San Jacinto Mountains, Riverside and San Diego Counties. Osborne Biological Consulting, Riverside

Osborne KH, Redak RE (2000) Microhabitat conditions associated with the distribution of post-diapause larvae of *Euphydryas editha quino* (Lepidoptera: Nymphalidae). Ann Entomol Soc Am 93:115–134

Papworth SK, Rist J, Coad L, Milner-Gulland EJ (2009) Evidence for shifting baseline syndrome in conservation. Conserv Lett 2:93–100

Philips SJ, Anderson RP, Schapire RE (2006) Maximum entropy modeling of species geographic distributions. Ecol Model 190:231–259

Pickens BA, Root KV (2009) Behavior as a tool for assessing a managed landscape: a case study of the Karner Blue butterfly. Landsc Ecol 24:243–251

Pickett STA, White PS (eds) (1985) The ecology of natural disturbance and patch dynamics. Academic, New York

Pratt GF, Stouthamer R (2008) The genetic relationships between the El Segundo Blues from Los Angeles County and Santa Barbara County. Department of Entomology, University of California, Riverside, Riverside

Pratt GF, Hein EW, Krofta DM (2001) Newly discovered populations and food plants extend the range of the endangered Quino Checkerspot butterfly, *Euphydryas editha quino* (Nymphalidae) in southern California. J Lepid Soc 55(4):169–171

Proctor J, Woodell SRJ (1975) The ecology of serpentine soils. Adv Ecol Res 9:255–366

Radchuk V, WallisDeVries MF, Schtickzelle N (2012) Spatially and financially explicit population viability analysis of *Maculinea alcon* in The Netherlands. PLoS ONE 7(6), e38684

Rodrigues JJS, Brown KS, Ruszczyk A (1993) Resources and conservation of Neotropical butterflies in urban forest fragments. Biol Conserv 64(1):3–9

Saccheri I, Kuussaari M, Kankare M, Vikman P, Fortelius W, Hanski I (1998) Inbreeding and extinction in a butterfly metapopulation. Nature 392(6675):491–494

Schoch D (2007) Rare butterfly is winging back. Los Angeles Times, July 9, 2007

Schtickzelle N, Choutt J, Goffart P, Fichefet V, Baguette M (2005) Metapopulation dynamics and conservation of the marsh fritillary butterfly: population viability analysis and management options for a critically endangered species in Western Europe. Biol Conserv 126(4):569–581

Schultz CB (1998) Dispersal behavior and its implications for reserve design in a rare Oregon butterfly. Conserv Biol 12(2):284–292

Schultz CB, Hammond PW (2003) Using population viability analysis to develop recovery criteria for endangered insects: case study of the Fender's blue butterfly. Conserv Biol 17(5):1372–1385

Scott JA (1986) The butterflies of North America: a natural history and field guide. Stanford University Press, Stanford

Shields O (1967) Hilltopping: an ecological study of summit congregation behavior of butterflies on a southern California hill. J Res Lepid 6(2):69–178

Singer MC, Thomas CD (1992) The difficulty of deducing behavior from resource use: an example from hilltopping in checkerspot butterflies. Am Nat 140(4):654–664

Snyder NF, Schmitt NJ (2002) California Condor (*Gymnogyps californianus*). The Birds of North America Online

Snyder N, Snyder H (2000) The California Condor, a saga of natural history and conservation. Academic, London

Soulé ME, Bolger DT, Alberts AC, Wright J, Sorice M, Hill S (1988) Reconstructed dynamics of rapid extinctions of chaparral-requiring birds in urban habitat islands. Conserv Biol 2(1):75–92

Sousa WP (1984) The role of natural disturbance in natural communities. Annu Rev Ecol Syst 15:353–391

Sutcliffe OL, Thomas CD (1996) Open corridors appear to facilitate dispersal by ringlet butterflies (*Aphantopus hyperantus*) between woodland clearings. Conserv Biol 10:1359–1365

Thomas Reid Associates (1999) San Bruno Mountain Habitat Conservation Plan 1998 activities report for endangered species permit PRT 2-9818. County of San Mateo, Palo Alto, California

Thorne F (1970) Habitat: *Euphydryas editha wrightii*. J Res Lepid 7(3):167–168

U.S. Fish and Wildlife Service (1976) Determination that six species of butterflies are endangered species. Fed Regist 41(106):22041–22044

U.S. Fish and Wildlife Service (1984) Recovery plan for three endangered species endemic to Antioch Dunes, California. U.S. Fish and Wildlife Service, Portland

U.S. Fish and Wildlife Service (1997) Endangered and threatened wildlife and plants; determination of endangered status for the Laguna Mountains skipper and Quino Checkerspot butterfly. Fed Regist 62(11):2313–2322

U.S. Fish and Wildlife Service (2003a) Draft recovery plan for Behren's silverspot butterfly (*Speyeria zerene behrensii*). U.S. Fish and Wildlife Service, Portland

U.S. Fish and Wildlife Service (2003b) Final recovery plan for the Quino Checkerspot butterfly (*Euphydryas editha quino*). U.S. Fish and Wildlife Service, Portland

U.S. Fish and Wildlife Service (2007) Lotis Blue butterfly (*Lycaeides argyrognomon lotis*) 5-year review: summary and evaluation. Arcata Fish and Wildlife Office, U.S. Fish and Wildlife Service, Arcata

United States Department of Agriculture (2005) Land management plan: part 2 Cleveland National Forest strategy. Forest Service, Pacific Southwest Region, R5-MB-077

Watts CH, Larivière M-C (2004) The importance of urban reserves for conserving beetle communities: a case study from New Zealand. J Insect Conserv 8:47–58

Wayne L, Weiss SB, Niederer C (2009) Recovery action plan for the Mission Blue butterfly (*Icaricia icarioides missionensis*) at Twin Peaks Natural Area. San Francisco Parks and Recreation Department and Creekside Center for Earth Observation, San Francisco

Weiss SB (1999) Cars, cows, and checkerspot butterflies: nitrogen deposition and management of nutrient-poor grasslands for a threatened species. Conserv Biol 13(6):1476–1486

Weiss SB, White RR, Murphy DD, Ehrlich PR (1987) Growth and dispersal of larvae of the checkerspot butterfly *Euphydryas editha*. Oikos 50:161–166

Weiss SB, Murphy DD, White RR (1988) Sun, slope, and butterflies: topographic determinants of habitat quality for *Euphydryas editha*. Ecology 69:1486–1496

Weiss SB, Niederer C, Wayne L, Wenerton K (2012) Progress report for the Mission Blue butterfly (*Icaricia icarioides missionensis*) at Twin Peaks Natural Area. Creekside Center for Earth Observation and San Francisco Recreation and Park Department

White RR (1974) Food plant defoliation and larval starvation of *Euphydryas editha*. Oecologia 14:307–315

Wielgus RB (2002) Minimum viable population and reserve sizes for naturally regulated grizzly bears in British Columbia. Biol Conserv 106:381–388

Zonneveld C (1991) Estimating death rates from transect counts. Ecol Entomol 16:115–121

Zonneveld C, Longcore T, Mulder C (2003) Optimal schemes to detect presence of insect species. Conserv Biol 17(2):476–487

MIX
Papier aus verantwortungsvollen Quellen
Paper from responsible sources
FSC® C105338

If you have any concerns about our products,
you can contact us on
ProductSafety@springernature.com

In case Publisher is established outside the EU,
the EU authorized representative is:
Springer Nature Customer Service Center GmbH
Europaplatz 3, 69115 Heidelberg, Germany

Printed by Libri Plureos GmbH
in Hamburg, Germany